Citizen-Consumers and Evolution

DOI: 10.1057/9781137276803

Also by Mikael Klintman

ECO-STANDARDS, PRODUCT LABELLING AND GREEN CONSUMERISM (*co-authored with Magnus Bostrum*)

DOI: 10.1057/9781137276803

palgrave▸pivot

Citizen-Consumers and Evolution: Reducing Environmental Harm through Our Social Motivation

Mikael Klintman

Lund University, Sweden

palgrave
macmillan

DOI: 10.1057/9781137276803

First published 2013 by
PALGRAVE MACMILLAN

Palgrave Macmillan in the UK is an imprint of Macmillan Publishers Limited, registered in England, company number 785998, of Houndmills, Basingstoke, Hampshire RG21 6XS.

Palgrave Macmillan in the US is a division of St Martin's Press LLC, 175 Fifth Avenue, New York, NY 10010.

Palgrave Macmillan is the global academic imprint of the above companies and has companies and representatives throughout the world.

Palgrave® and Macmillan® are registered trademarks in the United States, the United Kingdom, Europe and other countries.

ISBN: 978–1–137–27681–0 EPUB
ISBN: 978–1–137–27680–3 PDF
ISBN: 978–1–137–27679–7 Hardback

A catalogue record for this book is available from the British Library.

A catalog record for this book is available from the Library of Congress.

www.palgrave.com/pivot

DOI: 10.1057/9781137276803

Contents

DOI: 10.1057/9781137276803

List of Illustrations

Figures

Table

DOI: 10.1057/9781137276803

Acknowledgements

During the preparation of this book, I was fortunate to have the advice of several colleagues. Two contexts have been particularly important for this book: From their various disciplines, the twenty-five researchers and the referent group in the programme Governing Transitions Towards Low-Carbon Energy and Transport Systems for 2050 (LETS 2050) have helped me find much more interesting questions than I would have been able to develop myself. In the programme, I have collaborated particularly closely with the programme leaders, Lars J. Nilsson and Annica Kronsell, as well as with Björn Wendle, Jamil Khan, Johannes Stripple, Lena Hiselius, and Roger Hildingsson.

The reference group of my work package – Dan Melander, Jan Lindgren, Kerstin Heikenfeldt, Lars Jonsson, Maj-Lis Svärd, Stina Nilsson, and Therese Karlsson – have provided me with important insights about practical policy challenges and opportunities. LETS 2050 has been financed by Vinnova (The Swedish Governmental Agency for Innovation Systems), Naturvårdsverket (The Swedish Environmental Protection Agency), Energimyndigheten (The Swedish Energy Agency), and Trafikverket (The Swedish Transport Administration) from 2009 to 2013.

In the project 'Furthering Markets of Green and Ethical Products: the Role of Retailers', the participants – Oksana Mont (project leader), Beatrice Kogg, Matthias Lehner, Oksana Mont (project leader), and Olga Chkanikova – have helped me refine issues about the relationships between retailers and consumers. This project is part of the research programme Hållbar Butik (Sustainable Retailing), and has

DOI: 10.1057/9781137276803

been funded by Formas (Swedish Research Council for Environment, Agricultural Sciences and Spatial Planning) in collaboration with Handels Utvecklingsråd (The Swedish Retail and Wholesale Development Council) from 2010 to 2013.

Furthermore, I would like to acknowledge two organisations as important discussion forums that have influenced this book: the first is the International Sociological Association (ISA), the Research Committee on Environment and Society, RC24, where Magnus Boström and I have served as secretaries (2010–2014). The second is the European Sociological Association (ESA), the Research Network 12 on Environment and Society, where I have served as elected board member (2009–2013).

Finally, my thanks go to Jenny, Leo, Bruno, Fred, and Matti for all the insights they provide on a daily basis, and for being my main sources of motivation.

DOI: 10.1057/9781137276803

palgrave▸pivot

www.palgrave.com/pivot

DOI: 10.1057/9781137276803

1
Introduction

Abstract: *Citizen-consumers and motivation challenges to reduce environmental harm are introduced. An approach that takes motivation seriously, the chapter holds we must take into account its evolutionary basis. The ecological and the material motivation approaches are presented. Both ignore motivation that is evolutionarily rooted in social motivation.*

Klintman, Mikael. *Citizen-Consumers and Evolution: Reducing Environmental Harm through Our Social Motivation*. Basingstoke: Palgrave Macmillan, 2013.
DOI: 10.1057/9781137276803.

Hope is as stubborn as ever. Despite daily catastrophes in the media – effects of human-made and natural environmental disaster, social inequality, and unfair economic priorities – loud exclamations of hope are constantly heard. In the magazine *Wired*, the renaissance man and well-known film director David Lynch proclaims a new level of enlightenment as the recipe for a better world:

> Now, if you heighten [consciousness] in the individual, it leads to enlightenment – that's super-consciousness. If you heighten that in the world, it brings a higher collective consciousness. (David Lynch, in *Wired* [12], 2011)

Academic social thinkers call for different versions of cosmopolitanism. The sociologist Ulrich Beck, for instance, arguing for an 'analytical-empirical cosmopolitization', suggests that the world is moving towards a single community with shared morality, and with a global respect for a number of 'othernesses' [my term] concerning nature and culture. Beck describes cosmopolitization as follows:

> The most interesting thing is the fact that the reception of such media reporting creates an awareness that strangers in distant places are following the same events with the same fears and worries as oneself. Strangers become neighbors!. (Beck, 2011, p. 1350)
>
> No longer can we regard them as poor strangers, inviting us to be good and compassionate. Instead we must now understand them as partners in a common cause: both 'us' and 'them' tied together in the interests of survival, in the challenge of mastering global risks. (Beck, 2011, p. 1357)

Such high and almost spiritual hopes (albeit with dark overtones) are translated by governmental and non-governmental organizations from a structural orientation into calls at the level of agency, requiring and 'nudging' citizen-consumers to take a large share of the responsibility. People in many societies are confronted with imperatives that they should 'shop ethically', 'care for the environment', and 'think glocally'. Media, public agencies, NGOs, and 'progressive' companies issue these messages and instructions about virtually all sectors of society such as energy, food, travel, textiles, and furniture. Yet, despite very loud environmental concerns, largely among a growing middle and upper middle class in the richer countries (Eurobarometer, 2011), and despite the increased prevalence of eco-products and eco-services marked with various eco-labels and certificates, these messages and initiatives have had less than sufficiently positive results in several sectors during the last few decades

DOI: 10.1057/9781137276803

(Devinney, Auger, & Eckhardt, 2010). It is still the lower socio-economic classes in the South and North that make the substantial 'environmental efforts', because of their low levels of resource use. Whereas figures indicate that, for instance, CO_2 emissions have been slightly reduced in the North and increased in some countries in the South (OECD, 2011), this is not least in result of demands of consumer products *produced* in the South and exported to, as well as consumed in, the North. Still, consumer patterns are changing rapidly in the South, because of growing segments with increased economic resources. In its prognosis for the period 2011 to 2035, the International Energy Agency (IEA) assumes a 'world demand growth for all energy sources', and that 'the prospects for oil and gas demand hinge to a large degree on future policy decisions', such as subsidies directed at renewable energy (IEA, 2011, pp. 1–3). Similar prognoses and prescriptions could plausibly be made concerning several other natural resources and emission challenges.

Unless we subscribe to a fatalist worldview, a major part of these issues – and the potential for changing the trends – can be boiled down to human motivation, whether that of politicians, policy makers, industrial actors, NGOs, or citizen-consumers. This book focuses on the latter, albeit in close relation to the former groups.

1.1 Re-inviting the concept of motivation

To gain a thorough understanding of the gaps between what many people envision, what we claim to be important in life, and what we actually do and support, this book tries to revive the term *motivation* in the broader context of social sciences. In line with its Latin root, *movere* (to move), Reeve has defined motivation as 'those processes that give behaviour its energy and direction' (Reeve, 2005, p. 39).

Talk in terms of motivation has a very particular sound to some people, connoting an individual psychological, neo-classical economic, and/or traditional rational-choice approach, where the individual is perceived as motivated predominantly by maximizing economic utilities and minimizing deprivations of such utilities. Moreover, these motivations are typically conceived as fully conscious, in this neo-classical economic approach that social scientists often associate with the term motivation (Thaler & Sunstein, 2008).[1] However, as Brody convincingly pointed out decades ago, *most social scientific approaches* have an implicit notion or theory of motivation (Brody,

DOI: 10.1057/9781137276803

1983). Yet, scholars rarely dare to explicate it in motivational terms, perhaps from fear of seeming to adopt a traditional rational-choice perspective with excessive belief in full consumer liberty to consciously maximize 'utilities' with too limited an understanding of the role of institutions, norms, and rules as forces reducing such liberty (Turner, 1987). Alternatively, for a sociologist to talk in terms of motivation might give the impression of being unknowingly eclectic. Nevertheless, in the context of this book, I find motivation is the best term to refer to the root values and driving forces of citizen-consumers, However, this also requires a very clear use of the term in every chapter.

From the dawn of late modern environmental concern to the present day, two broad scholarly approaches have been dominant in analysing the societal challenges and opportunities for citizen-consumers. They are not necessarily tied to specific disciplines or policy areas, but can be found in writing, debates, and decision making within parts of several disciplines and policy areas. In fact, many, if not most, of us who are dedicated to social examination of environmental issues have at least made implications that fall under one or both of these approaches. The first one, which I call the ecological motivation approach,[2] understands reduced ecological harm as an actual or potential primary value among groups of, or among most, citizen-consumers. The second one, here called the material motivation approach, conceives incessant material accumulation as an – actually or potentially – primary value among citizen-consumers. The diagnosis of the ecological motivation approach is typically that people do not fully understand the gravity of global environmental problems and how these problems are tied to political and daily choices of citizen-consumers themselves. A major remedy is for people to become better informed about the urgency of environmental problems and about how environmentally harmful habits could be altered (Dietz & Stern, 2002; Nolan, 2010). The diagnosis of the material motivation approach is that the market economy is not yet adapted to taking 'negative externalities' into account (Steg & Vlek, 2009). According to this approach – unless it favours an invisible hand – the market should be adjusted so that citizen-consumers are given (mainly economic) incentives to reduce the damage they do and disincentives to increase such damage (Atkinson, 2009; Attari et al., 2009; *The Economist*, 2009). Interestingly, although the ecological and material approaches to environmental problems have often been treated as opposites, they have considerable similarities. Their faith in the potential of citizen-consumers – through reliable knowledge authorities – acting consciously,

DOI: 10.1057/9781137276803

strategically, and predictably unites the two perspectives, although the ecological motivation perspective sees reduced ecological harm as a possible primary value, whereas the economic one perceives reduced ecological harm as a possible means to satisfy what they believe to be the primary value of material accumulation.

1.2 Taking social motivation seriously

Despite seeming clear-cut, the ecological and material motivation approaches are often combined and mixed in academic work as well as in policy practice. Ecological motivation proponents, and some subscribers to the material motivation approach, combine the preceding prescriptions with calls for strengthened environmental regulation. Moreover, they are often mixed with a third dimension: the social dimension. When these dimensions are mixed, social scientific research on citizen-consumers is typically of the blank-slate character, whereby any aspect of culture is seen as equally significant to people's motivations and practices; in such research, eclectic holism is the principle from which conclusions are drawn. This is manifest in the common research finding that citizen-consumers, consumption, and green concern are 'complex' (Peattie, 2010, p. 195). This complexity may be described by listing everything that matters. The renowned environmental social psychologist Stern' s assessment of factors behind environmentally significant behaviour is the following:

1. 'Attitudinal factors, including norms, beliefs, and values', related to the environment, yet also non-attitudinal factors, for instance 'about frugality, luxury, waste, or the importance of spending time with family'.; 2. contextual forces – interpersonal influences, community expectations, advertising, governmental regulation, monetary incentives and costs, physical difficulty by technology or built environment; 3. personal capabilities (e.g., skills, knowledge, and resources of individuals and the organisations in which they may participate.; and 4. habit or routine. (Stern, 2000, p. 416–417)

This is a truly eclectic, holistic picture. Still, although a broad picture might be a useful starting point, this book maintains that studies ending with claims of complexity have not gone very far beyond our everyday, commonsensical view of the world. The reader is left wondering if everything really matters to an equal degree, and if nothing constitutes the basis for something else within this complexity. It is the task of scholars

DOI: 10.1057/9781137276803

to analyse – theoretically and empirically – to what extent certain factors may be more fundamental, or in other ways more important, than others; in this case, as drivers towards taking environmental considerations in daily life or not.

The most serious problem with the eclectic pictures of citizen-consumer motivation is their treatment of the social dimension as a supplement. Thus, it is fair to call them manifestations of a weak social motivation approach. 'We should not forget the social' is often heard among both researchers and policy practitioners involved in sustainability projects. The social becomes an additional factor, which may spice up projects of citizen-consumers and environmental concern (although it is assumed that the social dimension never reaches the same degree of importance as monetary saving or environmental morality). To a large extent dimension has so far been studied on equal terms with a wide range of factors: economic, practical, 'moral', and so forth. The underlying assumption when the social is treated as supplementary is typically that citizen-consumers (i.e., human beings) are blank slates on which any motivation may be equally strong; and why not add the social aspects to energy projects in order to make them more motivating for the participants? Although the previously mentioned 'remedies' are likely to continue to play a crucial part in environmental policy making, this book questions these suggested diagnoses and remedies. Consequently, this book is intended to provide a substantial challenge to the main accounts of scholarly works as well as policy practice concerning the 'green' and 'ethical' potential of citizen-consumers.

It is important to stress that extensive research has examined, highlighted, and theorized social parameters and their roles for people to change their daily practices. Social identity theory, for instance, contends an interplay between solidarity and competition, as a factor important to behaviour (Stets & Burke, 2000). With the Motivation-Opportunity-Ability Model, Ölander and Thøgersen (1995) portray the significance of (socially based) opportunity for practices such as structural and infrastructural opportunities. In their Theory of Reasoned Action, and subsequently in their Theory of Planned Behaviour, Ajzen & Fishbein include what they call 'the subjective norm' that refers to what people think that others will think about a certain action that they perform (1980). All these theories have been used as frameworks in studies of environmental behaviour. It is beyond the objective of this book to review this rich literature (for reviews, see Bechtel & Churchman, 2002). The scope of this book is narrower: to develop a social motivation approach

DOI: 10.1057/9781137276803

informed by evolutionary theory and to apply this approach to the roles of citizen-consumers in light of environmental harm.

The main argument of the book stems from a fundamental irony: The social science model previously sketched, with its blank-slate ontology and frequent treatment of the social as a supplement, has been labelled The Standard Social Science Model (SSSM; a term originally used by Barkow, Cosmides, & Tooby, 1992, p. 23). I contend that this model implies and entails a shallow understanding and treatment of the social dimension as weak. A much more thorough understanding and strengthening of the social dimension can be achieved, however, by allowing research within SSSM to be influenced by disciplines that have typically been perceived as its main enemy (a perception that has only partially been untrue): evolutionary theory on human nature, a perspective practised by certain schools of anthropology (Bentley, Earls, & O'Brien, 2011); the interface between evolutionary psychology, cognitive science, and neuropsychology (Pinker, 2009; Ploeger, 2010; Wright, 1995); and so forth. This book holds that it is necessary to make use of – and to go beyond – 'Darwin-free' social sciences to arrive at a new and more fundamental understanding of why the middle and upper classes and society's institutions, despite all the verbal concerns, knowledge, and economic resources, do not do more, and how the North can take more responsibility to reduce environmental harm. Although scholarly work conducted on the basis of SSSM may be highly valuable for its rich descriptions and analyses of the social, cultural, and organizational preconditions for citizen-consumers to be involved in environmental and social issues, the social sciences need to be made compatible with evolutionary theory. The latter provides well-founded knowledge about human and societal motivations, which in turn rest on biological driving forces without which social organization would be impossible.

It should be noted that there is nothing new about trying to marry sociologically oriented research with evolutionary theory (leaving aside the nightmare attempts labelled social Darwinism some hundred years ago; for an overview see Leonard, 2009), although new and serious attempts have so far been quite marginal (Freese, Li, & Wade, 2003; Ploeger, 2010). What is new is this book's attempt to make environmentally oriented social sciences focus on how people's roles as citizens and consumers, as environmental saviours or destroyers, can be informed by evolutionary theory of human nature. Moreover, this book is written with the ambition of elaborating on, and providing certain answers to, the reverse issue, which has rarely been given due attention. This issue concerns what research within the Standard

DOI: 10.1057/9781137276803

Social Science Model may contribute with, contributions that evolutionary scholarship has so far been unable to offer. As we shall see, the evolutionary perspective in itself typically leaves several wide knowledge gaps that 'conventional' social sciences with their focus on citizen-consumers are better suited to fill.

I argue that the intellectual 'detour' (via a brief search for human nature) turns out to entail a renewed acknowledgement of *the high significance and further potential of the social sciences* – albeit somewhat revised – for analysing preconditions for reduced environmental harm in our roles as citizen-consumers. Drawing upon social scientific research on consumption and environment, and informing it with an evolutionary outlook may, in my view, make analyses at least slightly less all-embracing (avoiding claims that economy, technology, law, culture, social processes, and politics all matter for citizen-consumers' habits and choices), whereas at the same time having the ambition of explaining the most important preconditions, concerns, and practices of citizen-consumers. Bringing human nature into social scientific analyses can help us say something about factors that belong to our primary orientations: human motivations. These factors have evolved over the hundreds of thousands of years of *Homo sapiens*, and the many traits that we share with other primates have been evolved over an even longer time. The anthropologist Gat presents data constituting scientific consensus that human nature has developed in the ancestral environment through an adaptation process to the way(s) of life in hunter-gatherer society:

> The hunter-gatherer way of life, while, of course, also evolving a great deal over the genus *Homo*'s two-million-year history, covers 99.5 percent of that history, and more than 90 percent of the history of the species *Homo sapiens sapiens*. (Gat, 2000, p. 84)

Accordingly, *Homo sapiens* has survived as a result of our adaptation to hunter-gatherer society, equipped with the motivations that had survival value in 'the ancestral environment' (Buss, 2009). In late modern society, we cannot erase these traits; we can never become the blank slates that many scholars in the social sciences still imply, blank slates on which any values can be written, thus overriding these human driving forces; such a Darwin-free shortcut, the evolutionary view contends, is particularly unrealistic on a broad societal level (Runciman, 2008). What citizen-consumers, groups, and organizations can do, however, is to use their modern reflexivity to decide to make active use of these basic human driving forces, and create policies and programmes that meet these forces in ways that also lead to reduced societal damage, for instance to the environment.

In contrast to early scholarly attempts to marry social sciences with evolutionary theory (social Darwinism, which was as scientifically erroneous as it was racist and by extension fascist), newer attempts have been far more productive (Leonard, 2009). The principal idea of the latter perspectives, typically founded on what has been called New Darwinism and sociobiology (Wilson, 2012), is that people around the globe are in all fundamental respects highly similar, some would say essentially identical, across cultures. This unity of human nature mainly evolved in hunter-gatherer society. As this sociobiological perspective (sometimes called evolutionary psychology) is growing, many comprehensive studies indicate that 'the noble savage' is a faulty myth, as is the myth of the civilized man of superior morality and intelligence (Lopreato, 1990).[3]

An illustrative example of empirical research using the new evolutionary perspective is the recent anthropological mapping of human universals. These are traits that can be found to a high degree in all societies, from contemporary societies resembling those of the ancestral environment to the most technologically advanced. What are found to be human universals can be assumed to be most closely tied to our driving forces, feelings, instincts, and motivations developed over hundreds and thousands of years as part of human nature:

> The reasoning is simple, whatever is constant through all human societies must be due to something that goes with people wherever they go; that would certainly include human nature (Brown, 2004, p. 51).

What do human universals reveal about the primary driving forces, the motivations of citizen-consumers, and the motivations that they share with people all throughout human history? When searching through the human universals index developed by Brown (2004), where cultures around the world have been studied – some being isolated from other cultures – I was intrigued to note that among some 200 universals, environmental concern is *not* represented, not even at the local level (which goes beyond concern for basic hygiene in the immediate proximity of one's primary group). This could possibly be explained with the contention that environmental concern might not have been crucial in the ancestral environment, where *Homo sapiens* adapted to and survived during the vast majority of his time on earth. Instead, being geographically mobile and flexible, rather than ecologically

DOI: 10.1057/9781137276803

concerned, might have been a trait more favoured by evolution to lead to survival and reproduction.

An apparent absence of a fundamental, ecological motivation in humans has also been noted and explained by, for instance, the renowned biologists Paul and Anne Ehrlich in *The Dominant Animal.* They forcefully argue that,

> our perceptual systems have evolved to encourage us to ignore [gradual wide-scale threats, such as] population growth, global warming, loss of biodiversity, and land degradation. (Ehrlich & Ehrlich, 2008, p. 127)

Note that these scholars do not claim that evolution has somehow *missed* wide-scale environmental issues when the battery of concerns has evolved in *Homo sapiens* so far; Ehrlich and Ehrlich go much further, by maintaining that our human systems 'have evolved *to encourage us to ignore* [my emphasis]' such threats. How could such an encouragement of our ancestors (genetically inherited by us today) ignoring wider, long-term incremental environmental threats be beneficial and successful? Because it helped those in the ancestral environment who were genetically equipped with this lack of such environmental worries, making them better able to focus on issues more crucial to their survival and reproduction: immediate threats of food shortages, dangerous animals, poisonous plants, providing safety and security for the young ones, and so forth.

Furthermore, a relevant factor for this book is that an individualist driving force towards unlimited accumulation of material wealth (separate from a drive to strengthen or maintain one's social position) does not appear among human universals either. To be sure, trade and material appreciation may all very well be human universals to judge from the lists provided by anthropologists. However, the idea of a *primary* motivation towards *unlimited material accumulation* among individuals, *without social considerations*, is not supported by research on absolute human universals. Other studies draw the same conclusion, albeit in different terms. For instance, Henrich and colleagues conducted a large-scale cross-cultural study in four continents and twelve countries, among nomadic herders, horticulturalists, and indigenous peoples, for instance in New Guinea, besides urban people from various corners of the world. Subjects were instructed to play various well-established games assumed to indicate the level propensity to strive for unlimited material accumulation. Their main conclusion was the following:

> We found, first, that the canonical model – based on [material] self-interest – fails in all of the societies studied. (Henrich et al., 2005, p. 795)

Similar conclusions have been drawn in more conventional studies of experimental economics (Boyd et al., 2003). Unlimited material accumulation of individuals separate from social considerations might not have been favourable for survival at the dawn of *Homo sapiens* in his original setting, possibly since the mobility, flexibility, and limited techniques for conserving food did not confer any evolutionary advantage on possible mutants with a strong drive towards such material motivation (at the expense of social cooperation). Such mutants would probably have lower chances of spreading their genes and these motivational traits to the next generation. This claim is supported by anthropologist Cronk among others. By going through an extensive number of studies, he has found that material accumulation and wealth, although highly admired in many societies, is neither fundamental or universal to Homo sapiens. Instead, 'achievement of success as it is locally defined [as wisdom, skills, material wealth or something else] is a good predictor both of an individual's status and of his[/her] reproductive success' (Cronk, 2005, p. 177). Moving back to the already mentioned anthropologist Brown, who rests on an evolutionary metatheory: What he finds to be a part of all cultures (he calls them 'absolute human universals') – along with music, aggressive males, and thumb sucking – are *a large number of social factors* such as cooperation, conflict, reciprocal altruism, and status struggles. Brown and a wide range of evolutionary-minded scholars emphasize the essentiality of social factors, despite the typically narrow scientific analysis of genes, organisms, or single individuals (Barkow, Cosmides & Tooby, 1992; Bentley, Earls & O'Brien, 2011; Pinker, 2009). Absolute human universals are arguably more deeply rooted in us, closer to our emotions, and to our conscious as well as subconscious parts of our motivation, than are practices and concerns that flourish in certain cultures while being absent or weak in others (Brown, 2004; Cournoyer, 2004; Gaspar, 2006; Hirsh & Dolderman, 2007). By treating social motivation as universal, this book endorses what could be called a strong social motivation approach.

Accordingly, the approaches contending either that ecological motivation or that material motivation is, or can become, culturally transformed into primary motivation ignore the essential driving forces of humanity. This problem is not a matter of philosophical hairsplitting. It is even fair to assume the following: To the extent that environmental information, policies, and practical projects rest on these two perspectives, this may be one of the reasons why more comprehensive reductions of environmental harm have not actually taken place. Or to reverse the matter, to the extent that environmental policy endeavours have borne fruit in long-term reduced

DOI: 10.1057/9781137276803

harm – through modified, sustained activities of citizen-consumers – the planned or unplanned ties of these projects to primary human motivation could arguably be a main reason. Consquently, an argument in this book is that the apparent absence of – or at least weakness of – ecological motivation and of unlimited accumulation of material goods in the pool of human universals, as well as the strong and multifaceted presence of social factors in human universals, constitutes an important, and far too seldom considered, basis for the challenges – and the handling of the challenges – to goals of greening of citizenship, consumerism, markets, and politics. By examining a wide array of studies on human universals, and comparing these with results from evolutionary psychology as well as behavioural economics and 'happiness research' (Bok, 2010), the factors shown in the accompanying figure emerge as most relevant to the understanding of citizen-consumers concerning environmental problems.[4] The factors are all part of what in this book is called social motivation.

Reflecting again on our ancestral environment: The evolution of a focus among hunter-gatherers on more direct threats and problems, and of a motivation for handling these through social collaboration (as well as social conflict) turned out to be much more productive than developing an all-encompassing concern for all kinds of possible threats (such as environmental ones dispersed in time and space). Moreover, the former-mentioned focus and social motivation of hunter-gatherers turned out to be much more productive than separate individuals being mainly concerned with how to accumulate their own material prosperity. Yet, as explained in subsequent chapters, this social motivation is to an important extent subconscious, instinctive, and emotive rather than based on fully conscious and purposeful reasoning.

Needless to say, the factors that are included in our social motivation are highly multifaceted, despite the simplistic illustration in Figure 1.1. Social

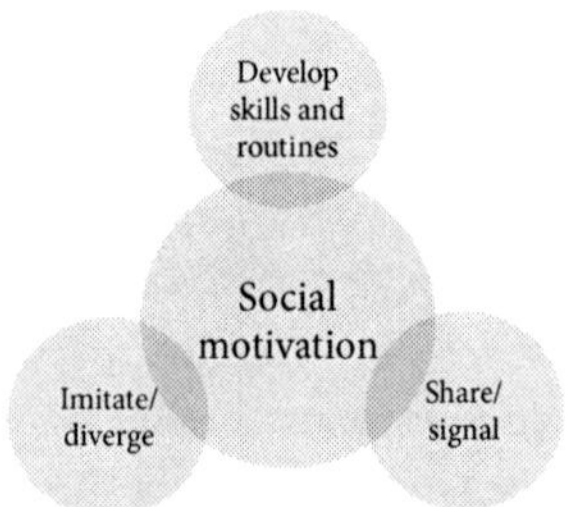

FIGURE 1.1 *Social motivation*

motivation is so diverse, partly subconscious, and culturally embedded that the preferences may be flexible within a wide range of factors of social positioning, collaboration, status struggles, likeability, and so forth. Already in the early 1980s, Brody showed convincingly that social motivation may very well be purposive and directional without being controlled by (conscious) cognition (Brody, 1983). Moreover, Brown, who was previously mentioned, expresses this diversity in the following way:

> in the social realm, universals include a division of labor, social groups, age grading, the family, kinship systems, ethnocentrism, play, exchange, cooperation, and reciprocity; in the behavioral realm, universals include aggression, gestures, gossip, and facial expression; in the realm of the mind, universals include emotions, dichotomous thinking, wariness around or fear of snakes, empathy, and psychological defense mechanism. (Brown, 2004, p. 47)

Furthermore, when environmental concerns become a strong norm in a group, *it may appear* as if environmental concern has turned into an overriding motivation. However, what has actually happened in such a case is that the group has channelled its group dynamics into environmental concern (as a means), thereby strengthening the social position and dynamics. Social relations are subject to primary, overriding motivation, but the social dynamics and cultural practices may be enormous for citizen-consumers in the efforts and struggles that this entails. Importantly, however, as multifaceted as the social dynamics and cultural practices are, they are neither arbitrary nor limitless, as the evolutionary perspective convincingly informs us (Alcock, 2003, p. 131). This should be kept in mind when analysing and developing schemes for less environmentally harmful practices of citizen-consumers. By being used as a positive side effect or means to satisfy the inherent social motivation of citizen-consumers and groups, reduction of environmental harm arguably has a great further potential.

This book uses theories of environmental sociology, sociology of consumption, theory of political consumerism, and political theory of deliberative democracy. The theory of economic rationality and ecological economics is also brought in to illustrate the two perspectives against which I try to argue. A parallel theme throughout the book concerns potentials and limits of taking social science literature that is in part social constructionist, and combining it with the essentialist literature on evolutionary theory. The way out of this potential incommensurability is to acknowledge that patterns of social constructions that we as sociologists analyse. For instance, that people frame certain

DOI: 10.1057/9781137276803

environmental issues as urgent but others as irrelevant, and that people frame certain groups as reliable participants in environmental decision making while excluding others as unreliable, constitute social patterns that in turn are tied to essentials of social interaction and of human nature.

The empirical basis of the book is case studies on citizen-consumers' environmental engagement – or lack of engagement – through their pecuniary activities, product use, and discursive sub-political practices. Data from various sectors and countries are used as examples to illustrate as well as to put the book's main thesis to test. The sectors include food, bioenergy, wind power, eco-taxes, mobility management, and various voluntary climate-gas-reducing policies directed towards citizen-consumers. The countries that are covered are especially Sweden and other Northern European countries. Moreover, in order to provide a broad international picture and to avoid extrapolating regional results as universal, the book makes use of extensive secondary data from other parts of the world.

In addition to analysis, the book discusses how to utilize the perspective of this book in the development of practical environmental projects and policies. To the extent that actors in the spheres of politics, market, and civil society are intellectually convinced that global environmental harm needs to be reduced in order to reduce risks of further deterioration of living conditions for humans, animals, and ecosystems, I claim that the three societal spheres need to collaborate in linking the necessary efforts to the more deeply rooted human universals. The most obvious ones are processes of social cooperation, group strengthening, and efforts to re-embed the consequences of harmful Northern practices to make them more visible. Another way of tying environmental efforts to human universals is to make the efforts part of less endearing human traits such as social status competition, social distance, and exclusion. Regardless of what human universals are chosen, it is crucial that policy practitioners, scholars, companies, NGOs, and citizen-consumers try to take accounts of human nature into consideration in this respect. Suggestions are given through brief examples of pioneer projects and further ideas for new ones. The basic challenge is how the social components of the human universals can be used by society as a whole, where the spheres of politics, markets, and civil society are involved. To a certain extent this is a daunting challenge that involves double reflexivity on the part of societal actors. Yet, with the combination of the social

DOI: 10.1057/9781137276803

motivation and the intellectual capacity located in the three spheres, it is well worth trying.

The book develops this thesis further by comparing it with the ecological motivation and material motivation approaches to citizen-consumers and their 'green' potential. The following two chapters examine the three approaches in more depth. In Chapter 2, this is done by comparing the approaches to see how they conceive of citizen-consumers concerning certain terms. What roles do people's values and social structures have for their practices? To what extent are citizen-consumers (subconsciously) reflexive, actively reflecting in their practices, and what role do habits play? Under what conditions are habits changeable? The question that overrides all of this is the following: In what respects are citizen-consumers who choose to take environmental issues into account rational, irrational, or even non-rational (a-rational)? Chapter 3 continues by analysing how the three approaches may explain the current or potential role of information and various knowledge authorities as bases for citizen-consumers to change lifestyles and engage in issues of environment. By developing the ancient dichotomy of Apollonian–Dionysian into Apollonian–Dionysian trust, I try to integrate social scientific and evolutionary perspectives to shed light on differences between the ecological, material, and social motivation approaches. My main argument here is that environmental projects, schemes, standards, and other voluntary instruments rely far too much on Apollonian trust. Apollonian environmental information that yields to individual reason, sanity, and self-restraint must give some of its space to the Dionysian dimension – social belonging and down-to-earth practice – in order to reach the primary motivation of citizen-consumers as *Homo sapiens*: social motivation.

The three subsequent chapters continue the preceding theme by highlighting three realms highly relevant to the roles of citizen-consumers in their environmental impacts (positive or negative; political or private). Chapter 4 deals with the lifestyle realm, how people's values and practices are related. Chapter 5 moves one step up and investigates what makes citizen-consumers ready or unready to initiate or accept changes of policy and planning in an environmental direction. Chapter 6 examines more general notions of broad societal change towards lower environmental impact and a post-consumerist society. To what extent have such notions taken the citizen-consumer into account as *Homo sapiens*? Chapter 7, the last chapter, draws important conclusions and discusses how policy practice and scholarly work on citizen-consumers could proceed, not least in order to fill the further knowledge gaps exposed by this book.

DOI: 10.1057/9781137276803

Notes

1 I only sketch the stereotypical view that some social scientists have of classical economics, that they associate with human motivation. In this book, I will try to illustrate a much more nuanced and practical view of economics.
2 The ecological motivation approach should be placed in a more general category, the moral motivation approach, the contention that any moral message could be written on the blank slate of human beings, and thus become intrinsically motivating to us, separate from other driving forces, such as the social motivation or our driving force of meeting basic material needs.
3 Although this may seem like a truism to some, certain self-righteous claims about differences in 'environmental concern' imply such ideas, if mainly concerning the relative sophistication of cultures.
4 These factors take place on top of essential driving force to satisfy basic, yet limited, needs of food, drink, security, clothing and shelter for oneself and one's kin.

DOI: 10.1057/9781137276803

2
Three Approaches

Abstract: *The ecological and material motivation approaches are placed in relation to rationality. Both approaches portray citizen-consumers as (actually or potentially) rational as in conscious, goal-rational, and focused on facts. Yet, citizen-consumers often act out of habit, instinct, and learn from imitation. Still, this behaviour is more 'socially rational'.*

Klintman, Mikael. *Citizen-Consumers and Evolution: Reducing Environmental Harm through Our Social Motivation*. Basingstoke: Palgrave Macmillan, 2013.
DOI: 10.1057/9781137276803.

The role of citizen has traditionally been one in which we are expected to look beyond individual direct interests. Engagement in social and political issues, struggle towards the common good, albeit partly through political conflict, are practices typically associated with citizen virtue. On the other hand, the role of consumer has traditionally been understood as the counterpole to the citizen role, although historians have shown many exceptions. As consumers, we have been assumed to act out of individual interest and utility maximization, for ourselves and our near ones. During the last two decades, this distinction has partly disappeared through analyses, policy practices, and campaigns as well as efforts by the general public shedding light on how consumption can be a way to reduce various ills in society, such as environmental harm (Trentmann, 2007). The term citizen-consumer refers to the hybrid role we all have today. This obviously raises a number of questions about what this role is, what it could become, how much people and nations vary in this role, and how citizen-consumers could have an impact on politics, policy making, and industry.

Academically, a large number of studies and disciplines examine the actual and potential role of citizen-consumers concerning environmental and social problems. The disciplines include social psychology, sociology, business economics, behavioural economics, ecological economics, market studies, political science, history of ideas, and ethnology. To make matters even more multifaceted, there are conflicting views across schools within each discipline.

Instead of summarizing each discipline and school of thought about citizen-consumers and environmental problems, this chapter elaborates on three themes that are necessary for any scholar or practitioner trying to understand citizen-consumers' current and potential role for societal change. These themes are values, reflexivity, and habits. The themes have been subject to the most important controversies in scholarly attempts at understanding all of us as citizen-consumers. In many cases, it is obvious that scholars or practitioners have not thought sufficiently about these themes, which has led to inconsistent or highly fragmented pictures of citizen-consumers. A claim in this chapter is that such inconsistencies are to a high degree because of social scientists typically not having integrated their perspectives with an explicit account of human nature and evolution.

This chapter places three themes in relation to a concept that serves as an umbrella, namely, the concept of *rationality*.

DOI: 10.1057/9781137276803

2.1 Citizen-consumers and rationality

Rationality is one of the most contested and misconceived concepts in scholarly work on citizen-consumers and environmental issues. As to the rationality concept in general, the debate seems to have been raging since the dawn of written philosophy:

> After at least 5,000 years of vigorous discourse, rationality remains a concept whose interpretation is susceptible to personal preference, idiosyncratic explication, and popular misunderstanding, and, therefore, has produced countless varieties of meaning. As a result, at the beginning of the twenty-first century, there is no universal agreement on what it means to be rational. A proliferation of meaning makes it necessary for responsible authors to say exactly what they mean by rationality. (Hammond, 2007, p. xii)

To make things simple, I offer the following working definition of rationality:

> actors or groups having, or with the potential to have, certain goal(s) for which they are motivated to use their resources (according to their hierarchy of preferences).

In this definition, it is perhaps more illuminating to note what is absent than what is present. First, this definition does not specify that *only one specific value content*, for instance material accumulation, should be more rational as means, end, or both in order to be called rational. For example, could citizen-consumers who consistently pay more for less environmentally harmful products or who struggle to make the government raise green taxes also be considered rational, even if it is financially costly to them? Second, this definition does not specify whether practices must be *conscious* to be rational or to what extent subconscious/tacit practices will do. Third, it does not specify whether the goal(s) or practices must be constantly *recalculated* within or across citizen-consumers to be called rational or to what extent habits would qua;ify as rational.

In contrast to the previous definition of rationality, the rationality of citizen-consumers (also called 'the general public') is often understood as inseparable from these three factors. Why is this? To my understanding, the rationality concept has been 'hijacked' by a very specific perspective within this material motivation approach. The rational citizen-consumer is typically portrayed as *Homo Economicus*, a classic figure in social thought. The notion, dating back to Adam Smith, is an understanding of each human being as motivated ultimately by economic self-interest, thus portraying humanity as consciously renegotiating their choices, using the best information and knowledge in a reflective manner, with the fixed preference of wealth maximization (for a positive description, see Binmore, 2005). Recent research in this tradition emphasizes the importance of economic factors as incentives or disincentives to comply with voluntary policy schemes. This can be expressed as the 'willingness to pay' for a certain environmental service or as economic incentives needed to use public transport, recycle, or to consume less electricity. Moreover, besides the emphasis on material values and individualism, the ecological motivation approach is highly consistent with several of these factors, which are treated as intrinsic to rationality. To be sure, very few scholars, policy practitioners, or corporate actors would subscribe to this caricature – in the economic or ecological version – if interviewed in depth about themselves or their peers. Yet we can see bits and pieces of the simplistic view as reflected in consumer policies, marketing, and in politicians' often overcautious attempts at bold political decisions about societal change – from local to international levels.

Although scholars who make claims about humans as rational decision makers (such as rational choice theorists) often do not make the preceding categorical claims, in the public debate nuanced claims are often translated sloppily into such unnuanced views. For instance, assertions about (subconscious) rationality that can be measured statistically, at the aggregate level (in large populations where clear patterns of rationality may be found) are translated in the media and public debate into conclusions about how *each individual* therefore must consciously think and strategically act. Thus, huge individual variation and cultural diversity are ignored, and the view of the motivations of citizen-consumers is distorted. Intriguingly, the reverse mistake is made by many social scientists. By conducting studies mainly from interviews and observations (avoiding numbers and statistics), these researchers acknowledge the obvious: Humans are not rational in the sense of consciously calculating, using all perfect knowledge and so on. Thus, these scholars fail to see rationality at the aggregate level, in larger groups of society, how seemingly altruistic practices may not at all be non-rational, but that such practices may be highly beneficial not just

DOI: 10.1057/9781137276803

to others but also to oneself. Voluntarily doing favours for others, giving to charity, and perhaps changing one's patterns of consumption may not only be ethical and environmentally sound, but also rational in several respects (without the individuals being aware of or calculating it). Axelrod's classical studies of how collaboration can be shown statistically to be rational as a non-zero-sum game convincingly demonstrates that a drive towards a degree of generosity and 'reciprocal altruism', also directed to others outside of the family, not only had survival value in the ancestral environment, but is also considered rational in modern society (Axelrod, 1997).

But why do good parts of social science conclude that no rationality processes are going on? Why embrace this cryptic notion of *a*rational citizen-consumers? The probable reason is that these parts of the social sciences have fully accepted the previously mentioned hijacking, the simplistic (and faulty) interpretation of rationality, although it makes this part of the social sciences erroneously criticize rationality as such. Moreover, they distance themselves from rationality claims because they are often seen as an obsolete part of sociology. Therefore, rationality is turned into a bad word in much of current social science, making any scholarly claim of people's rationality the main, sometimes their only, object of criticism. However, this is as unfair to the concept of rationality as it is to human nature.

To be sure, it is highly relevant to point towards *ir*rational traits of citizen-consumer activities in late modernity. When analysing citizen-consumer concerns in relation to environmentally hazardous practices, it is crucial to take into account the colossal societal changes that have taken place since all essential traits of humans were developed in our ancestral environment. Yet, to acknowledge irrational traits logically presupposes some kind of acknowledgment of the rationality of citizen-consumers. Several disciplines, from dynamic psychology[1] to anthropology and evolutionary psychology, point out the differences between the ancestral environment and current late modern society as the main root of several problems (which are most likely to have been less common in the ancestral environment); for instance, our pandemic levels of obesity and depression. To such problems one can add, at least partly, current environmental problems, in the sense that late modern citizen-consumers intellectually understand the problems of mass consumption but act according to motivations developed during times when environmental degradation, obesity, and extensive depression

DOI: 10.1057/9781137276803

had not yet become issues threatening human lives and reproduction. As Freese, Li, and Wade put it,

> Because our cognitive mechanisms are adapted for life in past environments, they cannot be assumed to produce evolutionarily optimal behaviors in the radically different environments of contemporary developed societies, and some theories propose to explain social phenomena as consequences of this mismatch. (Freese et al., 2003, p. 237)

This means that human practices typically perceived as 'irrational' could often be interpreted and understood as perfectly rational, if placed within the context of the ancestral environment. Nevertheless, Symons, among others, has shown that despite the intellectual capacities of humans, our minds cannot be expected to turn problems that are specific for advanced modern society into objects of our *primary motivations*, driving forces developed through adaptations to an ancestral environment with highly different problems (Symons, 1995). This misfit reasoning has been expanded further by Richerson and Boyd in their attempts to develop 'work-arounds', or ideas for how to organize a complex society consisting of members whose motivations were adapted to the ancestral environment (Richerson & Boyd, 1999).

These three ambiguities about rationality are fundamental to disputes about what citizen-consumers are, and could become, with regard to the challenges of handling environmental problems. This chapter continues by discussing these questions and giving examples in order to develop the conceptual framework.

2.2 Values as primary or secondary driving forces

Another term that has become mandatory in studies and practices concerning citizen-consumers is values. This is true regardless of the level of significance (from none to all) one attaches to values for environmentally related practice (Dunlap & York, 2008). What roles do values play for the monetary, use-oriented, and discursive practices of citizen-consumers? Should it be considered more rational to hold certain values as more vital than others? The concept of values is used in a wide range of disciplines, particularly in economics and psychology. Yet the context of citizen-consumer and environmental concern makes it necessary for us to go beyond the strictly economic and psychological perspectives of values, and move to broader social scientific perspectives. For instance,

DOI: 10.1057/9781137276803

this means that the rich development of value research in economics and psychology is not covered here.

Turning to the social sciences and the humanities, there are of course several definitions of values. Some definitions are more specific, for example, they contend that values 'guide selection or evaluation of behavior' (Schwartz & Bilsky, 1987, p. 551). For the purposes of this book, however, it is more useful to present a definition that leaves open the relationship between values and practice. Therefore, I have selected a definition where values are said to be

> interests, pleasures, likes, preferences, duties, moral obligations, desires, wants, needs, aversions and attractions, and many other modalities of selective orientation. (Pepper, 1958, p. 7)

To be sure, this definition is both broad and vague. Yet, at this point, this should not make us try to force values into a stricter format. What should be done, however, is to distinguish briefly between values and similar terms. First, values are different from attitudes in that the latter are more specific. For instance, a person's negative attitudes towards compromises in animal welfare might reflect her ethical or metaphysical values. Second, values are different from norms. Whereas the concrete practice of purchasing meat that is labelled organic can be a norm among a certain group of people, this might reflect their values. In contrast to values, attitudes are thus specific views whereas norms concern concrete practices. Although the previously mentioned distinctions do not turn values into something very clear cut, a way forward is to try to make use of this breadth of values when examining the role they play in issues of societal concerns. With this ambition, it is nevertheless possible to make the concept more manageable by distinguishing between various forms of values. A division into four value forms is useful, namely: ethical, aesthetic, metaphysical, and material value forms (Miegel & Johansson, 2002).

Let us exemplify value forms with issues of food and agriculture. The first form of values can be called *ethical values*. These are probably what first comes to mind as what is right or wrong (morally) about various agricultural and food-related processes in terms of environmental impact (locally and/or globally), the working conditions for farmers, or the welfare of animals. It is important to note that ethical values refer to issues that are both substantive (the impact of the actual products and production processes) and procedural (whether the decision-making processes of agricultural and food-related policies have been fair,

DOI: 10.1057/9781137276803

inclusive, and so on). The second form, *aesthetic values*, denotes the aspect of a product or production process being stimulating or not stimulating to the senses (when food tastes or looks appetizing, for instance). This meaning of aesthetic is the most obvious one. Moreover, influenced by sociology, one should emphasize a different – and even more relevant for this book's purposes – aspect of aesthetic values, where these refer to the value in terms of the self-expression, style, and identity of individuals and groups. Concepts such as 'eco-chic', 'green fashion', and 'organic food trends' partly imply aesthetic values. Moving to *metaphysical values*, these refer to positions and practices corresponding or not corresponding to various outlooks of life. Halal, Kosher, and vegetarianism based on outlooks of life are examples of manifestations of metaphysical values. Finally, *material values* have to do with conceptions of what choices of food products and production processes are economically or practically valuable, in this case, concerning societal values.

Two main questions emerge here: How are the values related to each other and what roles do values play? Are some values more 'rational' than others?

2.2.1 The ecological motivation approach

The ecological motivation approach assumes a (current or potential) direct link between motivation and environmentally sound practice. Various practical projects and policy schemes have been developed with ecological problems as the main frame. The hope is that the schemes will be accepted and used by citizen-consumers for *ecological* reasons (albeit in relation to other rationales such as money, time, and effort). The concepts of product boycotting as well as its counterpart, buycotting (e.g., choosing climate-labelled products) – both defined through green and ethical drivers – are frequently used as the operational definition of (monetary) political/green/ethical consumerism. Other, less explicit values often fall outside studies of voluntary schemes and the citizen-consumers that use these schemes. Lack of economic resources, for instance, lies behind a large part of people's limited negative impact on natural resources, through limited car use, energy use, avoidance of flight-transported food products, and so forth (Boström & Klintman, 2009).

The ecological motivation approach contends that ecological aspects of citizen-consumers' values are, or may become, a fundamental part of human nature. Ecological aspects of their values can accordingly be

DOI: 10.1057/9781137276803

dormant, However, they can and should be awakened through solid environmental information provided by trusted knowledge authorities. Alternatively (as in version A), the ecological motivation approach may imply a 'blank slate' on which the citizen-consumers may choose to write which values they would like to bring closest to their heart. If knowledge authorities give environmental and practical information about problems and how to handle them, the citizen-consumer can become motivated to write 'ecological concern' on her blank slate. These two versions of the ecological motivation approach have in common a view of ecological concern as a currently or potentially fundamental part of human motivation, in its own right, rather than being a potential, positive side effect of, or means towards, satisfying a more deeply rooted social motivation. In this approach lies a strong hope that citizen-consumers are, or may become, rational in the sense of deeply motivated and ecologically calculating, as regards how to change lifestyles, put pressure on market and politics, and go against group pressures and current materialist norms of mass consumerism. This approach contends that because environmental threats arguably constitute global threats, leaving materialist selfishness behind and acting for the global good is the only rational thing to do as a citizen-consumer. Please note that the ecological motivation approach may overlap the commonly separated perspective of 'small is beautiful' (the ideal of moving towards a simpler, less industrial society) and 'ecological modernization' (referring to the ideal of moving deeper into modernity, through greener technology and organizational innovations).

As heart-warming as the ecological motivation approach of citizen-consumers may seem, the approach is problematic in several respects. Here a few of the problems are mentioned briefly, whereas subsequent sections of the book cover the problems (and virtues) of this approach in greater depth. First, the data speak for themselves: A vast majority of citizen-consumers in the Global North are concerned or highly concerned with various environmental problems. They are highly aware of connections between their own daily lives and environmental and health-related consequences. Yet they (myself included, I must admit) do very little in practice to reduce environmental impacts or to support more comprehensive environmental policy measures. Information and knowledge are not enough. Ecological rationality, although everyone may understand what it means, does not run very deeply in most of us. It seems to remain in the intellectual realm of citizen-consumers and has

DOI: 10.1057/9781137276803

a very hard time evoking their deeper emotions. In more philosophical terms, one can say that the ecological motivation approach commits the 'naturalistic fallacy' (Moore, 1903). The fallacy is committed when inferring an *ought to* (e.g., that citizen-consumers should be motivated to act) from an *is* (environmental problems that citizen-consumers know about, if we allow ourselves to ignore here various knowledge uncertainties about environmental problems). In other words, there is something missing here between *is* and *ought*. Further on, I try to explain and fill in this missing link.

The following section, however, elucidates the role of values and rationality in the second approach, namely the material motivation approach.

2.2.2 The material motivation approach

As in the ecological motivation approach, the material motivation approach may be a fundamental part of human nature or the material motivation approach may imply a 'blank slate' on which the citizen-consumers can choose between which values they would like to bring closest to their heart. In a market-oriented, consumerist society (as in version B), people are highly inclined to write 'material values' in the largest letters on their blank slate. To give flesh and blood to this view, the material motivation approach often distances itself from the *Homo Economicus* perspective, by admitting that other aspects of ethical, aesthetic, metaphysical, and materialist values may be written as well on the blank slate, although in much smaller letters. In either situation, material motivation is a fundamental driving force. Yet, because the citizen-consumer and other actors and organizations in society may at the same time acknowledge the urgency of environmental problems, society can develop material incentives and disincentives that provide citizen-consumers with material benefits or costs in such ways that reduced environmental harm comes as a positive side effect of, or means towards, satisfying a more deeply rooted material motivation. In this approach, there often lies a view of material values as 'rational', in the sense of most beneficial to the citizen-consumer herself, at least in the short-term. The selfish, materialist rationality is accordingly difficult or impossible to hamper. According to Adam Smith, a founding father of classical economics, selfishness may also be beneficial to society as a whole. If everyone acted 'rationally', by seeking to satisfy their interests based on good

DOI: 10.1057/9781137276803

product and service information, individual and societal interests would converge (Smith, 1776/1982). But again, the material motivation approach is consistent with the view that society may develop policy schemes that make such rationality entail means or side effects through which negative externalities (such as ecological harm) are reduced. This means that the materialist motivation approach encompasses all perspectives that conceive material motivation as the most fundamental driving force of citizen-consumers, perspectives from neo-classical economists to scholars and practitioners who try to stimulate material values to overlap with ecological values. The material motivation approach includes a variety of views on humans – views that may refer to human nature, culture, or both.

Interestingly, the understanding of the rationality of the material motivation approach is not shared only across the previously mentioned perspectives. Scholars and practitioners in certain schools of thought where the material motivation approach is rejected – analytically as well as normatively – conceive rationality in the same way. In their illustrative book of daily life, *The Design of Everyday Life*, Shove and colleagues reject the view of the consumer as 'a rational actor', whom they describe as a consumer 'saving money, increasing property value' (Shove, Watson, Hand, & Ingram, 2008, p. 49)

Moreover, certain claims that use evolutionary reasoning fall within the material motivation approach. In the versions of the material motivation approach that discuss human nature, a common claim is reflected in the statement by Rees: 'Humans are Unsustainable by Nature' (Rees, 2010, p. 15). What does he, like many others echoing this claim, mean by this? Rees points to the fact that human nature, adapted to the ancestral environment of hunters and gatherers, has not been provided with a built-in 'off switch' concerning material accumulation and consumption. To be sure, there is obvious validity to parts of this claim. Excessive modern food consumption, with overweight and diabetes reaching pandemic levels, is a clear sign of the absence of a built-in off switch for the consumption of fat and sugar, substances essential for *Homo sapiens* to maximize the intakes of the unpredictable ancestral environment (Thaler & Sunstein, 2008).

Yet this argument is stretched too far, if it is concluded from the food case that *any type* of material accumulation and consumption by the individual is directly tied to human nature. A difference between excessive fat/sugar intake and general accumulation and consumption

DOI: 10.1057/9781137276803

of material goods is that excessive fat/sugar intake seems to resist many of the social norms about sound lifestyles, whereas the accumulation and consumption of (other) material goods still overlaps the social norms. The accumulation and consumption of automobiles, large homes, extensive travel, clothes, and gadgets are still, apart from some general and abstract criticism of 'mass consumerism', major signs of social status in modern mainstream culture. In the ancestral environment, unlimited material accumulation by individuals or subgroups at the expense of the physical needs of the rest of the band would have been practically impossible – partly due to the incessant mobility of these societies. Therefore, it is not plausible that such a trait would have developed in human nature as a trait separate from the social motivation. When we examine the human universals mapped out in various anthropological studies, we find a battery of social factors (social status struggles, collaboration, conflict, group distinctions, and so on) that are not dependent on material accumulation. Instead, material accumulation (above the essential physical needs), in many places seems to serve as *means* towards social status, belonging, and other parts of the social motivation, rather than as intrinsic ends. As regards practical implications, this distinction, where material accumulation is a means (among several) towards the social end(s) in human nature, gives some hope about the potential of citizen-consumers to reduce negative environmental impact (directly or through political support). If people from many groups and positions in society developing schemes and projects for reducing environmental harm could converge with the social motivation of people and groups, this would possibly be enough motivation; mass consumption far beyond our physical needs, if we made it diverge from social motivation, would possibly be enough to discourage it.

A more general criticism of the material motivation approach is that it implies an excessive belief in the power of economic incentives and disincentives to citizen-consumer in order to steer them in an ecologically sustainable direction. Economic (dis)incentives may certainly be useful for a period of time to stimulate alterations of specific routines. However, once the economic (dis)incentive is removed, the motivation to perform the ecologically sound practice may very well be removed as well. This follows logically from the material motivation approach, and it has been found in several empirical studies. Whitmarsh emphasizes the insufficiency of economic (dis)incentives. He brings nuance to the

DOI: 10.1057/9781137276803

picture by stressing the importance of combining economic (dis)incentives with other types of interventions:

> incentives should not solely be financial. Since a range of motivations (and barriers) underpins energy consumption and conservation, economic policies alone are insufficient to alter cherished and entrenched behaviours. Highlighting other tangible benefits, such as the health benefits of walking, for example, might offer one such means of encouraging alternatives to driving. The relative popularity of recycling highlights the need to provide facilities, such as kerb-side collection to facilitate pro-environmental lifestyles. (Whitmarsh, 2009, p. 21)

As the next section shows, other scholars criticize the material motivation approach (although they use other terms) by indicating how this approach often interprets the material motivation as the reason for citizen-consumers' changes of practice, whereas the reason lies somewhere else.

2.2.3 The social motivation approach

As the figure indicates, the social motivation approach acknowledges the multitude of values and value-based expressions of people as citizens as

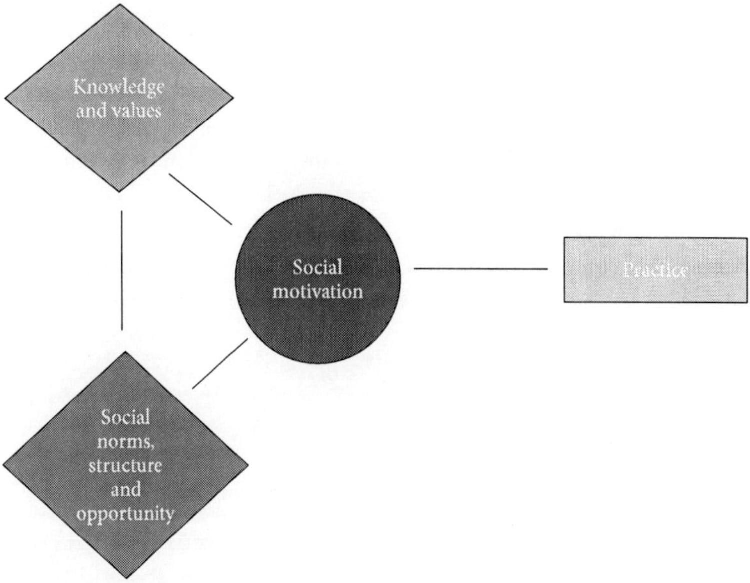

FIGURE 2.1 *Social motivation in context*

well as consumers. The social motivation can make much use of the social sciences, not least environmental sociology and sociology of consumption, which provide rich descriptions of how tastes, habits, views of what should be the responsibility of citizen-consumers, politicians, and corporations differ across groups, cultures, nations, and regions. Moreover, the social sciences give very useful insights into how values and norms differ within groups across sectors: What is considered a normal lifestyle or reasonable responsibility in terms of environmental impacts varies substantially across the sectors of, for instance, housing, transport, food, and energy. The social sciences also show, through in-depth studies of citizen-consumers, how the material motivation approach should not have a monopoly on defining rationality. A broadened rationality, where the full dynamic of values can be included, is more accurate than one where the citizen-consumer merely seeks maximum material value. All of this is elucidated by the social sciences and can be made fully consistent with the social motivation approach. Still, in contrast to parts of the ecological and material motivation approach, although values are flexible (admittedly constrained in part by social structures and norms), the content of the four value forms is not primary to us as humans. Instead, value contents are – very important – means to satisfy the overriding level, namely the social motivation. Differently stated, the value contents constitute a secondary driving force, whereas the social motivation is the primary driving force.

It may be useful, however, to repeat that whereas value content and its preferences are unfixed, such as the importance a group or society attaches to 'environmental responsibility', the existence of the *forms* of values, such as the four previously presented, are stable parts of human nature. For instance, everywhere and throughout the entire history of *Homo sapiens*, groups have developed distinctions between right and wrong. This is evidence that the ethical value form is stable. Darwin wrote the following about conscience, which can be seen as part of ethical values:

> Ultimately our moral sense or conscience becomes a highly complex sentiment, having its first origin in the social instincts, largely guided by the approbation of our fellow-men, ruled by reason, self-interest, and in later times by deep religious feelings, confirmed by instruction and habit, all combined, constitute our moral sense or conscience. (Darwin, 1879/2004, p. 157)

Whereas Darwin was probably thinking about conscience in relation to our direct, local or quasi-local social environment, it could today be broadened to sentiments driving interest organizations and Internet

DOI: 10.1057/9781137276803

communities to engage in wider issues, not least environmental ones. One should note, however, that the social interaction and limited anonymity strengthen such activities, since the 'instinct' that Darwin points to as the origin of conscience remains social. Other value forms are stable as well. Like the ethical value form, the metaphysical one constitutes human universals. Deliberations about our existence, where we come from and where we are going, are expressions of metaphysical value forms, stable parts of humanity (Newson & Richerson, 2009). The aesthetic value form is also part of all societies: having opinions about what is appealing to the senses and which is not, and appreciating evolutionary advantageous traits such as symmetry, scenery from safe, high positions, and so forth is also universal (Smith, 2005). Yet their content may vary substantially, not least as a means for marking group conformity or distinction. Aesthetic values are often used as social motivation as well as appreciation of skills such as handicraft, musical talent, and so forth. Similarly, the material value form – views of the value of goods and services – is a human universal, existing in all societies. Yet a key argument of the social motivation approach is that the universal material value form allows for a wide range of value content – from the most mass-consumerist ideal to anti-consumerist or 'post-consumerist' values.

Value content is established on the basis of social motivation, in combination with the basic driver of satisfying fundamental physical needs. To trace ethical values that reflect any intrinsic human motivations to protect the physical environment beyond one's immediate living area, motivations that operate independently of social motivations and social norms, would be to hope for too much of our ethical values. In that sense, Rees, whom I previously challenged for his material motivation approach, is correct when criticizing the view that knowledge about environmental deterioration may itself lead to ethical values that directly trigger citizen-consumers' motivation to change route (Rees, 2010, p. 20). Yet what is hampering such steps is not a fundamental material motivation. Rather, it is the fact that our modern institutions have not yet been modified to make reduced environmental harm and increased environmental engagement converge with social motivations.

Although mass consumerism and material accumulation in late modern society have so far been a fairly reliable means to strengthen and maintain social status (Hirsch, 1999), this does not mean that the value *content* favouring mass consumerism and material accumulation has become, or ever has been, a primary motivation in humans.[2]

DOI: 10.1057/9781137276803

More than a century ago, Thorsten Veblen coined the term 'conspicuous consumption', referring to the latent function of our consumer choices to show our status and socio-economic power. Veblen's perspective has been invoked in conjunction with green challenges in consumer society in general (Mitchell, 2001; Urry, 2010) and particularly with regard to wasteful practice (Druckman & Jackson, 2009). However, the Veblenian perspective has been largely absent in studies of voluntary green instruments, their preconditions, and possible improvements. But Veblen' s notions are valuable for understanding the latent social expression of green lifestyle practices with regard to voluntary instruments as well. Intriguingly, the conspicuous effect of consumption and use may probably entail two sides of a coin: (a) as wasteful practice, such as the choice of flight transport, in cases where neither time nor money is saved compared to train transport; or (b) as environmentally beneficial, by people of wealth favouring eco-labelled products (arguably eco-friendly), not least in cases where they are much more expensive than the comparable generic products.

In line with my application of Veblen to current environmental issues, Krause convincingly maintains that whereas modern culture has so far been favourable to consumption, consumption remains an instrument, a means, which may be changed into other means towards goals inherent in social motivation. Krause, for instance, singles out green consumption as a potentially less environmentally harmful means to achieve social status:

> More so than other forms of consumption, the consumption of positional goods would be impacted by a shift in norms, as it is motivated by a quest for social prestige. Alternatively, some green products can be seen as positional goods. They usually have a cost premium and are becoming increasingly socially admired. A strengthening of this trend would not lead to less consumption *per se*, but could lead to the consumption of products with reduced impact. (Krause, 2009, p. 289)

Similarly, Miller, in his book *Spent: Sex, Evolution, and Consumer Behavior*, envisions a 'shift from conspicuous waste to conspicuous precision', which he perceives as 'a gradual dematerialization of consumption' (Miller, 2009, p. 124). The word conspicuous, alluding to Veblen of course, indicates that there is an inevitable social game going on here. As I show in subsequent chapters of this book, social thinkers and philosophers have brought the notion of materialism and mass consumerism as means towards something else – sometimes too far – by using *happiness* as their point of departure, when sketching a society, institutions, and a way of life (with room for a plethora of lifestyles) where reduced environmental harm (a post-consumerist society) converges with a happier life.

2.3 Reflection versus reflexivity

Reflexivity is another term frequently used in the social sciences when trying to understand citizen-consumers in late modern society. As we shall see, reflexivity is an ambiguous term (although it does not have as enormous number of meanings as rationality). Perhaps it is the ambiguity of reflexivity that has generated its success as a commonly used term in social theory about late modern life; everyone may find something that fits their understanding of mass consumption, environmental and socially conscious citizen-consumers who nevertheless have more negative impact than those who do not care so much about grand words such as sustainable development.

For our purposes, it is most relevant to describe reflexivity as (a) two sides of reflexivity as awareness, and as (b) the subconscious, impulsive dimension.

Reflexivity, as analysed in scholarly works on 'reflexive modernity', points towards two opposing tendencies among citizen-consumers (as shown in a). The first is reflexivity in the sense of people perceiving themselves (as when watching themselves in the mirror), scrutinizing their own lifestyle, their environmental impact, aesthetic, ethics, and so forth. Giddens conceives this tendency towards reflexivity as a major characteristic of late modern society. In this society, individuals have particular resources (such as access to information and its technologies) enabling them to reflect critically on the environmental and health-related risks and harm that society – including themselves – causes in everyday life; harm that transcends the place and time in which they live (Ekberg, 2007; Giddens, 1990). Reflexivity here refers to the act of individual subjects directing awareness towards themselves, reflecting upon their own practices, preferences, and even the process of reflection itself. Beck, although agreeing with Giddens' basic claims, attempts to present the other side of the coin of increased awareness of one's own daily practices and their potential global consequences. Thus, Beck draws attention to an increased 'reflexive unawareness', namely people's

DOI: 10.1057/9781137276803

knowledge that they cannot know and cannot control the broader environmental or social outcomes of (in my terms) our monetary, use-oriented, and discursive practices as citizen-consumers (Beck, 1999). He maintains that our times are characterized by a 'moderate ethos of doubt and uncertainty' (Colombo, 2009, p. 739). Citizen-consumers are thus aware of the limits to their knowledge as well as the limits to the environmental promises of marketing and eco-programmes. As a remedy, Beck maintains that more transformative measures are needed at a higher societal level, where monetary and discursive consumerism may have an important role to play. This is a practice that has so far emerged from the bottom-up, through the demand to be able to 'do one's bit' being articulated and enabled by innovative strategies on the part of environmental NGOs or eco-entrepreneurs. Nevertheless, it is often the case that such instruments need to be further developed and refined on the basis of input from groups of citizen-consumers through Giddens' (1990) 'positive' as well as Beck's (1999) 'negative' notions of reflexivity. Despite their differences, both scholars, like most sociologists influenced by them, emphasize reflexivity as awareness and, implicitly, rationality as consciously reasoning about one's practices to hopefully act accordingly in order to reduce various kinds of harm. In fact, Giddens', Beck's, and their followers' view is very much in line with the ecological motivation approach; accordingly, increased information about the global consequences of people's lifestyles is (or can become) primary motivation for citizen-consumers; by conscious reasoning based on this information citizen-consumers may (it is hoped) become motivated to substantively change their lifestyles and put bottom-up pressure on politicians and corporations to move in a green direction.

Spaargaren and Oosterveer, greatly influenced by the reflexive modernity perspective, state that when considering the aspects of sustainability, with respect to products and services, in terms of geographical distance, traceability, and the social and ecological conditions of their production and distribution, we become aware of the massive impacts of globalization on production and consumption. It is because of the globalization of consumption that life-politics can be said to connect the personal to the planetary in a direct, unmediated way in the present phase of modernity (Spaargaren & Oosterveer, 2010, p. 1896).

They use the term 'lifestyle politics' to refer to the 'de-routinization' that they hope for, based upon the previously mentioned information and reflexivity.

DOI: 10.1057/9781137276803

> De-routinization of practices, so we argue, can (be made to) happen both at the level of the individual and at the level of social practices. The direct consequences of agents reflexively (re)considering their existing practices and narratives will, most of the times pertain both to the level of the individuals and to the practices involved. (Spaargaren & Oosterveer, 2010, p. 1896)

Accordingly, the resources for being reflective and self-reflexive give citizen-consumers the potential to become concerned – in word and deed – about the wider negative implications of their lifestyles for society and nature (Micheletti, 2003; Neilson, 2010). In other words, this view of citizen-consumers is one of people driven – currently or potentially – by conscious rationality, a rationality based on ecological motivation as a primary driving force. Such a view is not far from what has been frequently presented by scholars within practice theory when they discuss rationality per se, whereas practice-theory scholars criticize it when discussing citizen-consumers.

On the other hand, the practice-theory-oriented scholar Alan Warde, for instance, subscribes to an alternative view of citizen-consumers, whom he perceives as for the most part *a*rational (as shown in b). Warde makes the distinction between rational and more spontaneous (emotionally driven) consumers, very much favouring the latter view: 'Indeed, some shoppers are calculating and rational, others impulsive, addicted, entranced or seduced' (Warde & Martens, 2000, p. 167). When discussing cooking in the book *Consumption, Food and Taste*, he describes how some 'present cookery as a matter of rationality rather than practical judgment' (Warde, 1997, p. 157).

Warde and other practice-theory scholars highlight what is *truly* the other side of reflexivity, in contrast to Beck and Giddens, namely reflexivity as the subconscious, impulsive practices seemingly conducted through the human autonomous system. In my view, the importance of this other side is backed up by a wide range of evidence in behavioural and evolutionary psychology as well as neuroscience; this other side is arguably at least as important to understand and handle as the consciously decisional side if we as citizen-consumers are to reduce environmental and social harm, and to impact politics in this direction. In full agreement with recent sociological theories about citizen-consumers (for instance practice theory), researchers in sociobiology, evolutionary psychology (Gangestad & Simpson, 2007), and behavioural economics (Thaler & Sunstein, 2008) stress the enormous importance of subconscious human processes behind our daily practices as citizen-consumers.

DOI: 10.1057/9781137276803

This is further supported by the very well established distinction between System 1 and System 2 in social psychology, the former referring to conscious reasoning and the latter to subconscious aspects of human decision-making (see Pierre, 2011 for a good overview of dual systems). Nevertheless, although practice-theory scholars are highly convincing in pointing out the importance of this reflexivity, their view of this reflexivity as the opposite of rationality should not be left unchallenged. From an evolutionary perspective, 'practical judgment' (as Warde mentions in the previous quotation as the opposite of rationality) is – more often than not – rationality par excellence. Practical judgment is where our knowledge and competence are best suited for much of our daily decision-making. The title of an authoritative book on decision making, entitled *Better than Conscious?*, is one among several works that confirm this claim concerning many situations of daily life (Engel & Singer, 2008). It would be wrong, as some scholars in sociology and indeed also behavioural economics do, to ignore that subconscious, intuitive practices are often an important part of human rationality. Human practices, similar to behaviours of 'lower species', are often based on subconscious processes rather than on conscious reasoning. To satisfy one's social motivation, for instance, thus increasing the likelihood of maximizing inclusive fitness, might never be a goal that is conscious to a person in her daily activities; yet the evolutionary perspective teaches that the evolution of our brains is of such a kind that we behave (often subconsciously) based on this rationality (Baddeley, 2009; Pinker, 2002).

An example of the rationality of the subconscious can be seen in the way that nature has endowed us with a high talent for self-deception, probably mainly because self-deception is the most convincing basis for deceiving others that we are better than we are. This is very well established knowledge in evolutionary scholarship (R. Trivers, 2011). When citizen-consumers are asked how important it is that humans reduce the damage they do to the environment, most of us answer 'important' or 'very important' (Eurobarometer, 2011). This is neither a conscious nor a subconscious lie. Rather, it is a simplification and generalization that we deceive ourselves into subscribing (Devinney, Auger, & Eckhardt, 2010). Thus, we can honestly make this claim about our high environmental concern but a few minutes later drive away in our large car. Perhaps this is similar to all of the couples who make vows of eternal fidelity to each other; very few of us need to cross our fingers behind our backs at such moments. We thoroughly believe the other person, as well as ourselves,

DOI: 10.1057/9781137276803

despite our awareness of all the grim statistics on infidelity. People and animals incapable of any self-deception could not have been seen as honest and trustworthy to others, and therefore could not have become involved in relationships that ultimately could give them offspring (with a similar lack of ability for self-deception).

2.4 Habitual or constantly recalculated practices?

Related to reflexivity, but different in important respects is the issue of habits. Habits are puzzling. How do they fit into late modern society, particularly a society which social scientists have maintained to be a polarity of a traditional, routinized way of life characteristic of earlier rural society, *Gemeinschaft*? Perhaps this is why the social sciences, after a period of substantial research interests in habits in the early twentieth century, largely ignored it for several decades (Camic, 1986). As Camic pointed out in the *American Journal of Sociology* a quarter of a century ago,

> There is no article on habit in the International Encyclopedia of the Social Sciences, no place for it in recent indices of the major sociological journals, and no slot for it in the annual reviews and the standard textbook. (Camic, 1986, p. 1040)

When reading the classics one runs the risk of too simplistically translating dichotomies such as *Gemeinschaft–Gesellschaft* into 'traditional society (habit-based)'–'modern society' (reflective and decisional). Habit, tradition, and doing what we do because we have always done it has long been conceived as the very opposite of what people do in modern society. This might be a reason why sociology 'virtually dispensed with the concept' of habits for decades (Camic, 1986, p. 1040). Importantly, as Toennies pointed out much earlier (Toennies, 1887), the two concepts take place in a continuum; current society and the daily lives of citizen-consumers of today have both ideal types. As for neo-classical economics, its long avoidance of the concept of habits is more obvious:

> Econs [*Homo economicus*] never make an important decision without checking with their Reflective Systems. (Thaler & Sunstein, 2008, p. 22)

I maintain that it would be professional misconduct for a researcher trying to understand citizen-consumers to ignore habits. A citizen-consumer who to a large extent follows habits runs counter to the image

DOI: 10.1057/9781137276803

of *Homo economicus*, both as an analytical construct and as a normative one: how a competent citizen-consumer should act on the market.[3] Two important questions can be raised here. The first one is whether habits are necessarily reflexive in the sense of being 'automatic'; the second one is whether citizen-consumers' habits should be seen as the opposite of rational practices.

The first issue, about habits and degree of automation, is important to ask, because our environmentally harmful practices are often explained (by ourselves as citizen-consumers as well as by researchers) and excused as habits. If habits are conceived as conscious acts, this implies a higher degree of responsibility on the part of the citizen-consumer. If habits are subconscious and automatic, in the sense often described by behavioural economists, for instance, more extensive work would need to be done by the citizen-consumer to change habits, and until then the excuse would lie closer than in the case of habits as conscious practices. In my view, habits may be anything from subconscious to fully conscious practice, and it is very possible for citizen-consumers to reflect upon them in the way that Giddens, Beck, and Spaargaren previously indicate. One should not exaggerate the automatic dimensions of what goods and services citizen-consumers habitually purchase and use. As many studies and conventional wisdom indicate, habits may change, despite our human tendencies towards inertia and status quo bias. The social motivation approach, subscribed to by this book, hypothesizes that changes in habits, when they occur, are often socially motivated (consciously or subconsciously). Changes in habits may be driven by social motivations by moving in two directions: one towards social conformity and the other towards social distinction. Let us take an example. To change from having driven the car to work to riding one's bicycle to work is considered one of the more challenging changes of habits in developed countries. Although people often (consciously or subconsciously) find it more safe and comfortable to continue their habits, a change to cycling may be in line with their social motivation. Obviously, if others at our workplace, neighbourhood, or household ride their bike to work, our change of habits to cycling might well strengthen our social position in one or several of the groups, strengthen our group identity, and so forth. However, to be the first one in our group to change habits is an act of social distinction, which may be much more risky and energy-consuming. Perspiration, the condition of one's clothes, hairdo, and time use are only a few factors that may be affected by changing from driving

DOI: 10.1057/9781137276803

to cycling to work, particularly before the workplace has adapted by installing showers, and so forth. Still, being the first to change travel habits may also be particularly rewarding, if it is appreciated and imitated by others. It may raise one's status to that of a pioneer. This has been the case ever since the dawn of *Homo sapiens*; the risk of going against conformity may bring considerable social rewards if others appreciate the change of habits.

Nevertheless, habits remain challenging parts of human lifestyles, both to understand and to change. The reason for these challenges may become clearer if we discuss habits in relation to the second question, concerning rationality. Just as habits may be anything between subconscious and conscious acts, they may be anything between irrational and rational. What do I mean by this? If we start with habits as rational, I would strongly disagree with both the material motivation approach and the ecological motivation approach about the way they mainly present *the advantages* of a citizen-consumer reflecting actively and critically on all her practices. Although there are many problems with our habits – from routine prejudice against other groups of cultures to environmentally harmful routines – it is crucial to acknowledge the other and equally important side of habits, which is indicated by evolutionary theory: Habits are an essential part of humans, in traditional as well as in modern society, and for a 'rational' reason. Humans have limited energy and limited capacity to store information in the short-term memory (B. Schwartz, 2005). This necessitates a certain level of *blasé* attitude and routinization, not only in rural but at least as much in urban society. The sociologist Simmel pointed this out in a famous article about the metropolis at the turn of the previous century.

> If the unceasing external contact of numbers of persons in the city should be met by the same number of inner reactions as in the small town, in which one knows almost every person he meets and with each of whom he has a positive relationship, one would be completely atomized internally and would fall into an unthinkable mental condition. (Simmel, 1903, p. 106)

Routinization and habits help us make decisions in reasonable time, typically without having processed every single piece of data, and without conducting the perfect, extended cost-benefit analysis before every practice, which both the material and ecological motivation approaches would perceive as ideal. Evolutionary theory rejects the image of habits as irrational and constant re-evaluation as rational. Habits, like instincts,

DOI: 10.1057/9781137276803

have evolved over millions of years, and often serve our primary motivations better than conscious reasoning. As the social psychological theorist George Herbert Mead acknowledged early,

> Nor is all of human intelligence mental. Not only do our inherited and acquired habits exhibit manners which do not disclose mental operations but a great deal of direct influence lies outside of the processes ordinarily termed 'thinking'. (Mead, 1938, p. 68)

Furthermore, as has been previously pointed out about changes of habits, we tend to follow the habits of people in the groups we belong to or wish to belong to. Habits make us at least partially predictable as humans, a prerequisite for the stability of social collaboration and communities. Sociologists are eager to point out that habits are tied to various social structures, institutions, norms, physical conditions (such as concerning less environmentally harmful habits, recycling, vegetarian food, public transport, and so forth). Two levels are operating here: one of these concerns is physical/organizational preconditions and to what extent infrastructure, transport systems, the food chain, and the coordination between policy agencies are responsible for health and environment. For instance, cities that are developed in ways that facilitate lifestyles with low negative environmental impact. The other level concerns norms and culture regarding what should be considered a 'normal' modern way of life in your part of the world. The latter contributes to the reproduction of the former structures and vice versa.

2.5 Conclusions and discussion

The following is a summary of the main factors of the social motivation approach with regard to the rationality of citizen-consumers.

First of all, *values* are to a high extent behind citizen-consumers' purchases, use of products and services, and discursive engagement within ethical, aesthetic, metaphysical, and material forms of values. However, the fact that the citizen-consumer is not a *Homo economicus*, who only cares about maximizing economic self-interest, and instead may care more about environmental or moral aspects, does not mean that citizen-consumers have to be irrational or non-rational, as both classical economists and their most fervent critics in the social sciences would have it. Rather, the values on which citizen-consumers base their

DOI: 10.1057/9781137276803

monetary, use-oriented, and discursive practices are used – flexibly but not arbitrarily – on the basis of the overriding driving force evolved in adaptation to the ancestral environment of humankind: social motivation through group convergence as well as distinction. Second, *reflection (conscious) and reflexivity (conscious or subconscious)* are operating in dynamic ways among citizen-consumers, individually, in groups, and at all societal levels up to what Ulrich Beck calls cosmopolitization (see the introductory chapter of this book). Currently, policy makers, non-governmental organizations (NGOs), and social scientists typically place excessive emphasis on the novelty of late modern reflexivity (as enhanced consciousness and self-awareness) as a potentially intrinsic motivator for political and lifestyle-oriented change. Here evolutionary theory, not least mappings and interpretations of absolute human universals, reminds us of the subconscious level, and that alterations of citizen-consumer practices must take the social motivation into account, the drive towards collaboration, group dynamics, competition, and so forth. The rich palette that constitutes social motivation has developed as a main trait that has led to human adaptation and survival in the ancestral environment, and is in this respect 'rational'. Nevertheless, there are gaps between the ancestral environment and late modern society that may make certain aspects of social motivation 'irrational', not least concerning the environmental impact of mass consumerism. Still, whatever change society aspires towards, the aspects of social motivation will remain as end drivers that must be taken into account. Lifestyle-oriented change and changes to a renewed discursive/political engagement will have to function as means or side effects, not the other way around. Finally, habits could be seen as rational in at least three respects: (a) by having the citizen-consumer save time and energy; (b) by having the citizen-consumer share social institutions, routines, and traditions with others, thus strengthening the social bonds with her groups; and (c) by being possible, despite the challenges of human inertia, to reflect consciously on, often by comparing one's own practices with those of others, and ultimately to change, if a change is in line with social motivation.

In the following chapter, these two polarities are brought together then analysed with examples of citizen-consumers' roles and practices concerning reduced environmental harm. Chapter 3 contends that state agencies' and NGOs' environmental information and communication, a factor intimitely tied to rationality is expanded on, namely the role of

DOI: 10.1057/9781137276803

knowledge. I discuss society's preoccupation with a certain type of environmental knowledge, and society's assumption that citizen-consumers' trust in this knowledge will make them environmentally engaged. For people to be motivated to change their practices, I argue, another dimension of influence and trust needs to be brought in, a dimension typically overlooked when environmental issues are discussed.

Notes

1 (e.g., Freud, 1930).
2 *Homo sapiens* should not be confused with private firms, for which the primary motivation is indeed material accumulation.
3 Still, it should be mentioned that behavioural social psychology and economics are increasingly bringing in the role of habits to the economic and social sciences; for instance, this is indicated through the terms of 'status quo bias' and human inertia (Kahneman, Knetsch, & Thaler, 1991).

DOI: 10.1057/9781137276803

3
Apollonian and Dionysian Trust

Abstract: *Apollonian trust is the trust in the knowledge basis and problem-solving potential of, for instance, environmental prescriptions. Yet, Apollonian trust is not sufficient for motivating citizen-consumers. Dionysian trust is needed, so citizen-consumers' trust that eco-friendly practices will meet their social motivation. Schemes using celebrities, social marketing, and group-based interventions illustrate this.*

Klintman, Mikael. *Citizen-Consumers and Evolution: Reducing Environmental Harm through Our Social Motivation*. Basingstoke: Palgrave Macmillan, 2013.
DOI: 10.1057/9781137276803.

Current society is frequently described as a knowledge society, a society in which sufficient and relevant knowledge is – or should be – the basis for our practices. This ideal is heavily stressed when it comes to environmental and social problems, not least in regard to the marketplace. According to this ideal, policy makers involved in these issues should exercise 'evidence-based policy making'; companies should make environmental and social improvements based on the environmental and ethical audits that help to increase the companies' knowledge about the environmental and social implications of current and potential practices. Finally, the general public is asked by the previously mentioned actors and by non-governmental organizations to acquire knowledge of how to reduce negative environmental and social impact as citizens and consumers.

However, as mentioned in previous chapters, citizen-consumers are fairly well aware of some major negative impacts of our daily activities. Excessive car use, flights, food waste, and inefficient energy use are a few of the activities that we are aware of, and we know about their connection to pollution as well as the dramatic inequality of basic material resources globally. Moreover, many citizen-consumers have an idea of how to change daily routines to reduce such problems (Eurobarometer, 2011). Despite this, and although many of us claim to make use of this knowledge – which largely stems from information provided by public authorities, scientists, NGOs, and companies – energy use, for instance, is increasing, not just in developing countries with a growing middle class, but in developed countries as well (see Chapter 1).

This chapter raises some fundamental issues about information and knowledge authorities as a basis for citizen-consumers to change their purchases, use, and (sub)political involvement as implied in the notion of green political citizen-consumers. This chapter does this by questioning certain ways in which trust in knowledge authorities is typically portrayed, namely as trust in the cognitive scientific competence as well as in the tendency to have various stakeholders participate in these cognitive, knowledge-acquiring endeavours (Mumford & Gray, 2010). As Bellaby notes, trust is not a clear-cut concept with a singular meaning. There are a plethora of definitions and uses about the concept of trust. A core notion of trust that constitute this chapter's point of departure is that it denotes reliance on another organization, group, or individual 'to act in one's interests' (Bellaby, 2010, p. 2625). As we will see, there are different notions as to what these interests might be.

DOI: 10.1057/9781137276803

3.1 Evolutionary theory and trust

In order to elaborate on the social motivation approach as a tool for empirical analysis of citizen-consumers, it would be very useful to have a few concepts that could float above the stiff distinctions between 'conventional' social sciences and evolutionary theory, between human nature and culture, and so forth. While preparing to write this book, my idea was to go through the most well-known dualities in social thinking and see if they can be used in the subsequent analysis of this book. The assumption was that the classical social thinkers often had a less distant relationship to evolutionary thinking than have many contemporary social scientists. These were the results:

Tönnies is well-known for having developed one of the most classical dichotomies, namely that between *Gemeinschaft* (community) and *Gesellschaft* (society), which is also mentioned earlier in this book.[1] This distinction – particularly as a continuum – is still very powerful for comparing a traditional rural society with a modern urban society and everything in between. Moreover, it is often used in analyses of and between subcultures. Yet, it is best suited for *spatially* oriented analyses of social life, whereas citizen-consumers and environmental implications need to be detached from the spatial dimension in this book. Another classical distinction comes from Charles H. Cooley, that of primary and secondary groups. This distinction refers to groups with which we have face-to-face association and actively socialize versus groups that are instrumental, impersonal, and anonymous. This distinction partly resonates with the ancestral environment (a primary-group society to which *Homo sapiens* is adapted) versus modern urban society (where secondary groups have become more numerous). However, this distinction does not do full justice to the subtle dynamics of citizen-consumer influence by many different subcultures and groups within their secondary groups as well as by media and advertising, in particular their effects on lifestyles and consumption. Weber's distinction between bureaucracy and charisma might lie closer to the subject of this book than the previously mentioned distinctions. This distinction refers to leadership, something that is crucial when discussing the influence, impact, and power of citizen-consumers as well as how motivated citizen-consumers may become as a result of various types of 'leadership' (from environmental agencies to peers in lifestyle issues). However, the bureaucracy–charisma dichotomy is too closely connected to the stiff

DOI: 10.1057/9781137276803

rationality–irrationality separation that evolutionary theory has helped to overcome. As we see in Chapters 4 and 5 of this book, following the charismatic leadership of marketing, advertising, and friends should be acknowledged for its rational dimension to the extent that it is based on efforts (conscious or subconscious) to strengthen or maintain one's social position. Although well aware of Weber' s highly sophisticated analysis of forms of rationality, there are more straightforward ways to develop useful concepts for this book other than to go into depth on Weberian analysis.

What about the distinctions from evolutionary theory and from its main disciplines? Evolutionary theory has been far less active than the conventional social sciences in developing social dichotomies and distinctions. This may have to do partly with the fact that the main 'project' of disciplines based on (Neo-Darwinist) evolutionary theory – sociobiology and evolutionary psychology, for instance – has aimed at studying similarities between cultures, races, classes, ways of life, and between groups in history and in contemporary society. The point of departure of evolutionary minded scholars of human society is typically that the social sciences are (overly) pre-occupied with differences across groups, social categories, and classes. From an evolutionary perspective the similarities are much more prevalent, important (and interesting) to understand, not least as a knowledge basis for social change. Although evolutionary research has made impressive scientific advances studying human cultures, advances entailing benefits from a humanitarian perspective, and by rejecting arbitrary social categorization (i.e., racial), this book's theme of citizen-consumers and environmental impact cannot be developed by ignoring the stark differences between cultures in terms of causing environmental impact and suffering from environmental impact. This means that scholarly work on this theme must take into account what research using an evolutionary framework often ignores, namely *differences* that take place in the field between the individual and mass consumerism. The evolutionary-oriented psychologist Geoffrey Miller admits this problematic gap that is typical within his research field.

> These sciences [political science, sociology, and media studies] have realized through decades of research that they can't jump straight from individual psychology to mass culture [...] The shaping of [...] ideas, tastes, norms, habits [...] by social power systems is exactly what the social sciences study. (Miller, 2009, p. 50)

Unable to find a dichotomy from social sciences or from evolutionary theory useful for this book's purpose, I have had to consult the ancient Greek thinkers as a point of departure for reconstructing the term 'trust'. As it turns out, the Greek distinction between Apollonian and Dionysian forms appear to be the most suitable point of departure. Whereas this distinction is best-known through the interpretation of Nietzsche, I use the terms rather freely with respect to Nietzsche' s view and instead rely more on ancient uses. I argue that two distinctions are needed in order to include, but also move beyond, the previously mentioned view of trust as a possible basis for changed daily practices. One distinction should be made between what I call Apollonian and Dionysian trust, and another distinction between simple and reflective trust. With these distinctions, I try to give nuance to the concept of trust, particularly with regard to 'relevant information' and information givers about current and potential environmentally relevant changes in routines as citizen-consumers. Importantly, this criticism is not merely directed against the ecological and material motivation approaches. Whereas this book endorses several claims within the perspective of reflexive modernization, the reflexive modernization perspective nevertheless has shortcomings in its overly cognitive, fact-oriented, and Apollonian approach to reflexivity.

By *Apollonian trust*, I refer to trust based on the substantive quality of the data and of the analyses on which recommendations are grounded. This book's notion of Apollonian trust is obviously rooted in the character of the Greek god Apollo, representing order, structure, discipline, and reason (Coolidge, 1941). Consequently, Apollonian trust is founded on the ideal of reason, individuality, sanity, balance, self-restraint, and on the hopeful belief that there are optimal and reachable results that should serve as the basis for the trust of, in this case, citizen-consumers, thus constituting the basis for environmental adaptation of these actors. We can conclude that trust in actors and organizations believed to provide optimal and healthy solutions is the core of Apollonian trust; Apollo is the god of medicine (Kourvetaris, 1997, p. 230). Importantly, Apollonian trust need not merely refer to scientifically founded knowledge, although parts of it should. In the light of recently broadened conceptions of knowledge, denoting that a broad range of 'stakeholders' and organizations should be part of its acquisition, beyond science and scientists, Apollonian trust refers to trust in actors and organizations based on their perceived performance in producing and presenting substantive, 'relevant' knowledge.

DOI: 10.1057/9781137276803

In my construction of the concept of *Dionysian trust*, it is crucial to stress its basis in, among other things, evolutionary theory. To be sure, the most frequently mentioned characteristics of Dionysus, the Greek god of wine, have to do with 'irrational' traits such as intoxication, ecstasy, subcultures, and wild dance, 'experienced always by the individuals in a group' (Kourvetaris, 1997, p. 231; cf. Nietzsche, 1872/2010). Yet, based on this book's idea of the social motivation as the evolutionary and overriding drive of humans, the drive that this entails – in view of its ties to reproduction, social status, belonging, protection, and so forth – is not necessarily 'irrational', but rather instinctive, sometimes subconscious, but nevertheless strategic. A citizen-consumer holds Dionysian trust when she senses – consciously or subconsciously – that allowing herself to adopt the prescription of a group – at micro, meso, or macro levels – or becoming involved in altering these prescriptions is in line with her social motivation. Thus, what has developed in humans as instincts, feelings, or spontaneity is closely tied to our social motivation: to do well in the social game. Much of what may be perceived as irrational is, from an evolutionary perspective fully, or partly, in line with our social values.

In isolation, the concepts of Apollonian and Dionysian trust say little about authority and participation. Both of them may evoke connotations of top-down authority; the Apollonian in its partial closeness to the ideal of scientific authority and the Dionysian in the images it may give us of individuals submerged by charismatic authority in ecstatic group exercises. Yet, in modern democracies, not least concerning imperatives issued to citizen-consumers in the market and civil realms, these concepts need to be placed in relation to another distinction, namely that of *simple and reflective trust*. This distinction has been developed by Magnus Boström and myself in two previous texts (Klintman & Boström, 2008; Boström & Klintman, 2011), where we analyse eco-labels. There we describe *simple trust* as the trust in knowledge authority that need not be questioned. Experts are most likely to have the best access to the relevant knowledge, not least concerning environmental problems, how to deal with them, and what priorities are most reasonable. Simple trust is a type of trust where citizen-consumers take the views of knowledge authorities literally and do not compare them to their own or other actors' experiences. On the other hand, *reflective trust*, which is a term inspired by Giddens' 'active trust' (Giddens, 1991), is a trust more in line with current ideas of reflexive modernity, knowledge democracy, and so forth. In our terms, reflective trust is analogous to the stereotype of the young adult's ideal trust in her parents (after the child's simple trust and the adolescent's categorical mistrust), namely the trust that the parents' claims may be fallible but are still worth a serious comparison with other knowledge claims, in sum generating a knowledge base that all parties may agree upon. In more formal writing, reflective trust is a more advanced level of trust in which consumers and other stakeholders acknowledge the fallibility, ideological diversity, and political compromises of environmental policies. *Reflective trust* is a trust that the standards can be improved, and that consumers and a wide range of stakeholders are needed in these processes of continuous modification, as individuals and as members of organizations (Boström & Klintman, 2011, p. 8).

The following sections go into more depth with the concepts of Apollonian and Dionysian trust, and their relevance for understanding citizen-consumers with regard to environmental problems.

3.2 Apollonian simple trust

The ideal of Apollonian trust that was introduced at the end of the previous chapter, with regard to citizen-consumers concerning environmental issues, is that of citizen-consumers unreflectedly trusting information or policies from authorities. Because authorities (at universities, in governments, NGOs, and companies in an 'objective' setting) have the helicopter view, and through their advanced access to facts and knowledge, they should be able to best assess what citizen-consumers and other groups of actors should do to reduce environmental harm. This ideal includes an outlook about how citizen-consumers can become motivated to change their daily routines and sub-political practices. Steg and Vlek illustrate the underlying notion clearly, when describing informational strategies:

> informational strategies can be aimed to increase actors' knowledge so as to heighten their awareness of environmental problems and of the environmental impacts of their behaviour, and/or to increase their knowledge of behavioural alternatives and their pros and cons. It is assumed that new knowledge results in changes in attitudes, which in turn will affect behaviour. (Steg & Vlek, 2009, p. 313)

Authorities, as previously mentioned, may develop their prescriptions and policies in a measured, balanced, and thoughtful way, not least

DOI: 10.1057/9781137276803

through structures of checks and balances that are in place in academia, politics, and the public debate. Such structures have strong connotations of Apollonian principles. Coolidge describes the Apollonian character not only as 'shaped', but also as individualized (Coolidge, 1941, p. 454). This points towards Apollonian simple trust as a trust of single individuals in the knowledge competence of authorities, a trust that turns into motivation to follow their prescriptions.

The ecological as well as material motivation approaches include this view, which can be found in a wide range of academic and practical literature on citizen-consumers and environmental adaptation. The whole industry of environmental auditing, monitoring, and certification typically puts its hope in a simple Apollonian trust (although recent developments include a more dialogue-oriented development of such schemes as discussed in Section 3.3). As Paterson and Stripple claim,

> The social study of accounting showed how such practices enabled the invention of 'calculative selves' and 'calculative spaces' that enclosed individuals within 'calculative regimes'. (Paterson & Stripple, 2010, p. 359)

Assumptions of the ecological (sometimes in combination with the economic) motivation approach are often considered the *raison d'être* for lifecycle analyses, eco-labels standards, labels, audits, and certificates. For instance, Heinonen and colleagues formulate their hypothesis as follows:

> Our hypothesis is that a compilation of indicators for a sustainable information society would help the ecological modernisation process utilise the power of the information age and thus improve our knowledge and lives. (Heinonen, Jokinen, & Kaivooja, 2001, p. 328)

The main challenge that the Apollonian simple trust ideal contends is that of how citizen-consumers can gain substantive knowledge about the ecological, economic, and practical factors; such knowledge accordingly entails increased motivation to adapt one's daily activities and subpolitics to these rationalities. This view of a linear relationship between knowledge and changes in citizen-consumers' practices is often called 'the knowledge deficit model' (Sturgis & Allum, 2004).

This model has received much criticism lately, particularly from the evolutionary-based research perspectives. As to scientific claims about climate change as an urgent threat, the amount of data and information supporting this stance are far larger and uniting main parts of the

DOI: 10.1057/9781137276803

scientific community far more than does climate scepticism. Nevertheless, an ambitious study published in *Nature* on people in the United States and their concerns about climate change show that people with the highest scientific literacy and capacity for technical reasoning were not among the most concerned ones. Instead, people highly literate in scientific and technical issues constituted the most culturally polarized category. Rather than using their knowledge and skills to support the position strongly elevated as 'the best available knowledge', by being highly concerned about climate change, many of these people used their scientific knowledge and skills to question and criticize the best available knowledge. The authors of this article draw the following conclusions:

> This result suggests that public divisions over climate change stem not from the public's incomprehension of science but from a distinctive conflict of interest: between the personal interest individuals have in forming beliefs in line with those held by others with whom they share close ties and the collective one they all share in making use of the best available science to promote common welfare. (Kahan et al., 2012, p. 1)

People's level of concern about climate change was based rather on their social motivation than on their scientific knowledge. It should be noted that people were divided in the study based on which of two 'worldviews' they subscribe to. Some of the knowledgeable people identify themselves with others with a more industry-promoting and individualist worldview; they turned out to be less concerned about climate change. Others among the knowledgeable ones identified themselves with a more egalitarian, communitarian worldview; they turned out to translate the mainstream scientific community's alarms into a high concern, just like their peers. As to the conflict of interests that the authors refer to in the previous quote, the interest in forming beliefs in line with their peers, as well as the interest in promoting common welfare through best available knowledge, are very close to the concepts presented later in this chapter's discussions about Dionysian and Apollonian trust.

There are more counter-intuitive lessons to be learnt about the relation between knowledge, values, and practices. Interestingly, although the concept of the knowledge deficit model has mainly been used to criticize simplistic versions of the same model (such as in the biotechnology sector; Sturgis, Cooper, & Fife-schaw, 2005), organizations opposed to new or 'risky' technology use the very same knowledge deficit model, yet by arguing that the more knowledge people have of such

DOI: 10.1057/9781137276803

technologies, the stronger their *opposition* (Marris, 2001). As to the idea of increasing, substantive knowledge about ecological damage increasing our willingness to reduce that damage, a frustration about scientifically or economically illiterate citizen-consumers is implied in much work in environmental behaviour, policy, and economics. For instance, some researchers on public opinion about using carbon capture storage to reduce the risk of climate change refer to public 'pseudo opinions' and 'non-attitudes' to describe initial local public opposition to this technique (Malone, Dooley, & Bradbury, 2010, p. 419). Their argument contends that if citizen-consumers' knowledge about ecological and economic benefits were to increase with this technique, their propensity to support it would increase.

Another area where the ecological and economic goal principles are frequently seen is household-related energy efficiency. For instance, Steg emphasizes how personalized, tailored information about energy saving options can be beneficial for households by addressing an individual's environmental and economic concerns, which implicitly motivates substantial changes in routines (Steg & Vlek, 2009).

Even if researchers agree that a basic ecological, economic, and practical understanding of environmental problems and the role of our daily practices is an important foundation for change, a common view among researchers is that information campaigns, in general, can hardly lead to practice changes (Abrahamse, Steg, Vlek, & Rothengatter, 2005). The gap between knowledge and practice in our daily lives is everything from a classic philosophical issue to a practical, empirical issue, and is also highly acknowledged by actors and organizations that provide environmental information and prescriptions to citizen-consumers. Section 1.3 mentioned the philosophical claim that 'is' should be separated from 'ought', in order to avoid the 'naturalistic fallacy'.

For our present purposes, as regards environmental auditing and policy, the field of industrial ecology (IE) works strongly on the basis of the ecological goal-principle. Nevertheless, it has been noted that the challenge of motivating changes among consumers and producers is in some ways the Achilles heel of industrial ecology. Despite over a decade of detailed analyses of the lifecycle impacts of different products, a deep understanding of the trade-offs between different goods and services, and vast improvements in tools to design more sustainable products, we still know very little about what actually motivates change within firms or consumers (Holzer, 2010). And perhaps more problematically, evidence is mounting that the gains of IE – such as eco-efficiency improvements in specific products – are being outpaced by increases in overall consumption levels around the world (O'Rourke, 2005, p. 116; Andrews, 2000).

However, there are notions of how this tension can be resolved. The main solution, or rather the way to handle this complexity (as there is no 'solution') prescribed by the social sciences and increasingly by certain governmental authorities, is typically *transparency, dialogue, and participation.* This chapter covers transparency, whereas participation is dealt with in depth in Chapter 5.

For instance, Ulrich Beck, the founder of the thesis of risk society, maintains that the disclosure and visibility of environmental information on products and services offer much promise (Beck, 2000). In analyses of 'political consumption', Micheletti and colleagues call for increased 'visibility' and 'transparency' as the means of enhancing green and ethically oriented activity by citizen-consumers in the market realm (Micheletti, Stolle, & Follesdal, 2003). According to this view, the main challenge is how to make unsensed 'abstract' risks and production processes that are usually separated from citizen-consumers' daily lives visible risks and processes. By extension, this view contends that the visibility of the environmental risks will in itself trigger consumer reflection and changes in daily routines, such as boycotts, buycotts, and more discursive protest action through Internet networking and so forth. The visibility will awake or stimulate our ecological values (ecological goal), with or without economic values, in turn leading to rational action to reduce the harm, in combination with, or separately from, economic calculation. Many researchers assume that this new visibility will stimulate more open public debates, in turn strengthening the empowerment of citizen-consumers (Crane, 2005).

However, in light of many policymakers' and scholars' calls for, in my terms, Apollonian *reflective* trust discussed below, it is intriguing to note the one-dimensional, quantitative character of the prescription of 'increased transparency', despite their often sophisticated theoretical elaboration of concepts such as reflexive modernity and deliberative democracy.

To give the buzzword *transparency* nuances that correspond to the reflective and reflexive schools of thought, I have, together with Magnus Boström, elsewhere developed the idea of *transparency layers* to go beyond the one-dimensional use of transparency (Klintman & Boström, 2008). This theoretical tool is developed further here.

DOI: 10.1057/9781137276803

The typical calls for increased transparency towards citizen-consumers in green issues take place, in our terms, through *the layer of simple, mediated transparency*. Elucidated here are issues such as, 'What do the green labels or certificates stand for?' Many environmental and ethical issues are obviously impossible to perceive by looking at the product. In the terms used in reflexive modernity theory, production processes are disembedded from the everyday life of citizen-consumers. The green and ethical information schemes and their expert developers can, if we apply Innis's depiction, function as 'an artificial body, or exosomatic organ that supervenes upon and penetrates the "natural equipment" with which we are endowed at birth' (2002, pp. 131–132). Simple, mediated transparency refers to substance: environmental criteria, thresholds for toxic chemicals, banned and permitted substances, salaries of farmers in the South, and so forth.

An important and probably never-ending debate is taking place between those who favour simplicity and those who favour complexity in the environmental and ethical information provided to citizen-consumers. On the one hand, there is the basic notion among economists of the ideal market situation, where consumers make their choices based on perfect knowledge of product characteristics, price comparisons, and – some would add – negative externalities, such as environmental or social harm involved in various phases of the product cycle. On the other hand, there is a worry about burdening consumers with 'information overload', and a worry that 'too much' information will make the products more expensive and lead to false or irrelevant implications about the uniqueness of one product compared to another (Horne, 2009). Moreover, concerns have been raised among sustainable-development actors that a lot of information about the environmental impact of a practice is often falsely portrayed, or erroneously assumed, as being equal to 'good environmental record' (Klintman & Boström, 2008).

The prescriptive framing 'more transparency' has so far been won by proponents of a larger quantity of substantive information addressed to citizen-consumers. The typical solution presented for the simplicity of eco-standards is that they should be supplemented with more detailed, technical information about the actual reductions of environmental and social harm promised by this 'alternative' product or service. Such substantive transparency is often called for in the light of market worries about green inflation and green washing.

DOI: 10.1057/9781137276803

One obvious reason for the challenges in developing Apollonian simple trust is that the knowledge and prescriptions given to citizen-consumers cannot always be the right ones. Given the constant fallibility of environmental and economic knowledge about changes in the activities of citizen-consumers, there have to be times when the prescriptions given are not the most useful ones. The risk of *simple* Apollonian trust is that it may turn into a blind *mistrust* in the authorities and their environmental prescriptions.

BOX 3.1 *Climate labelling: simplified or detailed information?*

Climate labelling may range from simplified categorical statements, such as 'climate neutral' or 'climate friendly', to intricate presentations of various types of climate gases associated with the product. The climate gas implications, or the environmental impact as a whole of products, are at least as complex to calculate as are the climate gas emissions calculated in carbon offsetting. Critics of climate labelling (simple or complex) hold that the feedback given to consumers through labels only reflects a rather modest – sometimes mainly symbolic – part of the total climate impact of the product. For instance, the mode of transport that the consumer uses on her way to and from the store is arguably more important for climate gas emissions than is the transport from farm to store. Also, how the consumer uses the products, how she disposes of it, or how much – in the case of food – is thrown away instead of eaten, is arguably more important than the mode of pre-consumer transportation (Engelhaupt, 2008). Thus, some would claim that the labelling schemes need to be supplemented by much more far-reaching consumer education and discussion that covers the whole product chain and the wider patterns of our daily lives (Klintman & Boström, 2011).

A large body of literature, not least in sociology and political science, has raised this challenge into a critique of the very principle of Apollonian simple trust (although this term has never been used). I call this critique and alternative view the ideal of Apollonian reflective trust.

DOI: 10.1057/9781137276803

3.3 Apollonian reflective trust

In a modern society with green and ethical empowerment and with responsibility expected of citizen-consumers, there is a fundamental tension between the notion of the importance of transparency and dialogue with 'reflective and reflexive' citizen-consumers in the light of all the knowledge uncertainty and ideological diversity (Beck, 1999; Giddens, 1990; Macnaghten, 2003) versus the increased prevalence of expert-based, standardized audits, eco-labels, and other relevant information (Power, 1999). Such schemes are largely carried out by various types of experts working in governments, NGOs, trade organizations, and private firms. The schemes become meaningless unless they are to some extent trusted (Power, 1999). Thus, a partial Apollonian simple trust remains necessary in the advanced 'audit society' in which citizen-consumers operate. In contrast to the ecological goal and economic goal perspectives, this book argues that it is unrealistic to hope for citizen-consumers who make full use of the ecological and economic information about products and services, making environmentally sound and/or economically optimal decisions as citizens and consumers. The complex calculations required to assess the relative ecological impacts of various goods, services, and technologies in terms of climate gas emissions and wider environmental impact are most often impossible for citizen-consumers to do (Taylor-Gooby, 2006).

Yet, as we have argued elsewhere (Klintman & Boström, 2008), the simple, mediated transparency geared towards an Apollonian simple trust is rarely sufficient in itself. The acknowledgement in theories of reflexive modernization about knowledge uncertainty and ideological diversity, also among scientific experts, necessitates a further layer of transparency. This is a transparency that exposes the decision-making procedures as well as some of the technical details. By revealing which organizations have been part of the criteria setting, and by indicating the political, strategic, and scientifically influenced process that leads to the inclusion of some criteria (such as chemical additives) and the exclusion of others (such as climate gas emission), this transparency layer, which we have called *transparency of negotiation*, can promote a more responsible, albeit not fully reflective, consumerism. According to this view, there is rarely *one* optimal solution, in terms of one policy or one set of changes among citizen-consumers, to solve problems of climate change, chemical pollution, reductions of biodiversity, and so forth. Or

DOI: 10.1057/9781137276803

at least there is no way of knowing for certain; 'knowledge authorities' – including scientists – are never neutral, and (Apollonian) trust of the public, at least in a simple sense, is therefore difficult or impossible to achieve. Transparency of negotiation enables citizen-consumers to acknowledge that information about a product or service, in terms of environmental outcomes, should rarely be seen as a reflection of 'truth' or 'what is *really* ecological' (Klintman, 2002). This layer is based on the interdependence between substantive factors (substances, thresholds, environmental outcomes) and procedural factors (strategies, interests, political processes). To make such interdependence transparent – for instance whether an eco-label is controlled by a NGO or by the retail chain itself, can in certain cases be more relevant for citizen-consumers than highly technical data.[2]

Still, the analysis of those propagating for measures to increase Apollonian reflective trust ends as a rule with the recipe for this type of trust, disregarding the fact that the step towards active engagement on the part of citizen-consumers must go beyond Apollonian reflective trust. For instance, Dore and Lebel give the following prescription for 'public legitimacy':

> Ultimately, public legitimacy of the wide range of state or non-state actors is based on the establishment and maintenance of public trust via transparency, accountability, responsible practice and competence. (Dore & Lebel, 2010, p. 132)

Despite its very welcome acknowledgement of the social and political components of environmental knowledge, those proposing transparency of negotiations, not least among scholars of reflexive modernization and deliberative democracy, should note that it is still chiefly occupied with cognitive dimensions, as if these were the fundamental source of motivation for ecologically concerned citizen-consumers. Thus, the question remains Apollonian, in the ideals of both simple and reflexive (Apollonian) trust: how can citizen-consumers become more knowledgeable about environmental issues, their connections to lifestyles, and sub-politics? This knowledge, either only substantive or also procedural knowledge, is assumed by both Apollonian camps, by its cognitive force and quality, to lead citizen-consumers to become engaged in reducing ecological damage through modification of their lifestyles and in changes in political support. Even if Apollonian trust, both simple and reflective, have their obvious necessary places in any

DOI: 10.1057/9781137276803

society with environmental challenges. Therefore, the statistics on citizen-consumers' negative environmental impact and modest environmental engagement – along with much research in sociology of consumption in public understanding or science and in the evolution of human values – raise questions that move beyond the preconditions for Apollonian trust. For instance, without any reference to evolutionary perspectives, the sociologist Elisabeth Shove accurately points out that

> Academic analysis on the identities of consumers, mirrors broader policy discourses that tend to focus overwhelmingly on information, awareness and individual consumer choice. (Shove, 2003, chapter 1)

What is needed for us as citizen-consumers to increase our environmental engagement and reduce our global harm more significantly? To elaborate on this question, we need to broaden the perspective to include Dionysian trust and its preconditions.

3.4 Dionysian trust

In this book, the notion of Dionysian trust with regard to how citizen-consumers handle environmental issues is based on the claim that more fundamental values and drives are at play among citizen-consumers than ecological, economic, and practical ones. With modern society's strong focus on the distribution of wealth, money, and materialities, subsequently accompanied by greater reflexivity about the distribution of environmental and health-related risks (Beck, 1986/1992), the main prescriptive framings in the literature and policy making have 'consequently' been to stimulate ecological goals/values and broaden economic goals/values, not least among citizen-consumers. However, the notion of Dionysian trust acknowledges that neither ecological nor economic values are overriding. Moreover, this evolutionary perspective accepts that humans are not blank slates where any values can be inscribed and become overriding, intrinsic, and fundamental – at least not beyond the margin. Instead, we must search beneath to acknowledge human universals. In such endeavours, we are joined by scholars in neuroscience, cognitive science, behavioural economics, and recent developments in sociology, as a reaction to a previous excessive focus on Apollonian factors: not least ecological and economic values, as a basis for acceptance and engagement of citizen-consumers.

DOI: 10.1057/9781137276803

To be sure, the lack of Apollonian trust – simple or reflective – is a central theme in science and technology studies about how the public perceives various authorities in regard to environmental risks. In many places, the Apollonian trust is low, for instance in the United Kingdom and in the United states, concerning governmental environmental agencies. Yet, coming from the Nordic countries, where the Apollonian trust in governmental agencies is generally higher, as with the Swedish Environmental Protection Agency, Chemicals Inspectorate, National Board for Consumer Policies, and so forth, the question remains why more people do not move from daily car use to public transport or why they increase their meat consumption.

Here it is obvious that Apollonian trust may be necessary but not sufficient. More effective for supporting pro-environmental household practice appears to be provision of information about how other people in a similar, and slightly higher, social position behave (Schultz, Nolan, Cialdini, Goldstein, & Griskevicius, 2007).

BOX 3.2 *The example of smoking*

A comparison can be made with smoking: Why is smoking, despite worldwide information about its hazards, still increasing in some regions, particularly among young people? What motivation may be operating beneath the values of keeping oneself and others healthy, and saving money for security and enjoyment? It has even been shown that labels such as 'smoking kills' may work as a trigger for some people to start smoking or to smoke more. Slovic (2001) uses the immense success of cigarette advertising as an example of the need to go beyond, or beneath, trust in scientific health claims (and by extension environmental claims). If we try an explanation in terms of Dionysian trust, a reason for the success could be that many people may hold an Apollonian trust in the health claims, but a stronger Dionysian trust in the providers and models of the cigarette ads, in addition to their own smoking peers. Instead of downgrading Dionysian trust into an 'irrational' trust, we should be open to analysing the positive aspects that such trust may embed. Perhaps it could be that the ads make many people associate smoking with strengthened social status, inclusion in intimate conversation with attractive people, and so forth. Perhaps such social gains are actually the case in many places. From an

DOI: 10.1057/9781137276803

evolutionary perspective such advantages could be far more valuable – not least by pointing towards increased chances of finding an attractive partner – than the governmental agencies' threat that smokers will lose a decade or two from their lives; that is, smokers as a collective, whereas some individuals will not. In this example, the social motivation may consciously or subconsciously drive people to smoke, unless governments, planners, employers, and so forth – together with groups among the general public – try to develop alternatives where social motivation may be met to a higher extent among non-smokers than among smokers. To the extent that governmental agencies are permitted to be involved in 'campaigns' to affect public opinion, they may have much to learn from marketing (as I will discuss in the section on social marketing). For instance, Nyborg and colleagues (2006) have shown convincing results about how advertising and marketing may affect how we think that people in our surroundings act, something that ought to be highly relevant to the environmental engagement of citizen-consumers.

3.4.1 Celebrities towards Dionysian trust

One type of attempt at raising the interest of citizen-consumers in ecologically beneficial lifestyle change has been to focus on celebrities who are involved in environmental efforts. For example, Leonardo DiCaprio has taken part in a campaign organized by the organization Global Green USA, followed by his purchases of 'green' hybrid sports cars. Celebrities are certainly of constant interest to the ordinary citizen-consumer. Is it not fair to say that all of us occasionally play with the thought of living a certain celebrity's life? This is an integral part of the human capacity for empathy and being interested in how other people are likely to act in various situations, an ability of great evolutionary value. Yet there are several problems with using celebrities to motivate us to eco-adapt our daily lives. Aside from the potentially suggestive power of people that we admire and perhaps also are attracted to, for most of us it is only possible to identify partly with celebrities. Everyone knows celebrities lead an entirely different kind of life from most of us, probably highly inconsistent in terms of environmental impact. Their eco-chicness may make us try to copy part of this aesthetic, yet with the risk of it being a fashion and fad, which makes Boykoff and Goodman call this role of celebrities 'conspicuous redemption' (2009). The newspaper headline 'Eco campaigner Sting flies solo on private jet – twice in one day' (*Daily Mail*, 30 June 2008), which appeared some years ago, is an illustration of this concept. Still, when used in a realistic way, invitations of celebrities to increase Dionysian trust may very well work. The anti-littering campaign 'Don't mess with Texas' has been successful during its many years of activities, particularly because of celebrity participation. In the first six years of the campaign, the amount of visible roadside litter was reduced by 72 percent (Thaler & Sunstein, 2008, pp. 60–61). The celebrities selected for this campaign have often been those who the main litterers admire such as Dallas Cowboys as well as other Texan athletes or music stars. The unifying allusion to Texan culture and identity is a further factor strengthening Dionysian trust. Finally, as opposed to the high risk of 'conspicuous redemption' in the case of Sting and other celebrities with a wide moral rhetoric about saving the planet, many would find it more realistic that Texan superstars do not throw litter on the roadside.

Instead of rejecting the idea of celebrities as ecological role models, one could also argue that a more comprehensive norm change must begin with fashions, fads, niches, and so forth, which in turn may be transformed into more solid societal norms and new 'regimes' (Geels & Schot, 2007). For the social motivation approach, aesthetics and fashion are in essence the very opposite of the temporary, and in the case of fashion shallow phenomena of social life. Instead, viewed as a social form and process, it concerns highly constant and fundamental parts of human nature as instruments for social distinction and attraction.

3.4.2 Social marketing towards Dionysian trust

The preceding example about smoking mentions that governments and NGOs may have much to learn from the marketing realm, even from the cigarette companies. From this thought, a type of marketing has evolved that is in line with social motivation, which may include issues of public health, reduced crime, and reduced environmental harm. This marketing is labelled *social marketing*, and it has been developed for use by governmental agencies, non-governmental organizations, and any other kind of group that has a social motivation going beyond financial profit (Andreasen, 2005). The term social marketing was coined by

DOI: 10.1057/9781137276803

Kotler and Zaltman in 1971 as the application of marketing techniques to achieve some kind of social change. In a later text, Kotler and colleagues define social marketing as the use of marketing principles and techniques to influence a target audience to voluntarily accept, reject, modify, or abandon a practice for the benefit of individuals, groups, or society as a whole (Kotler, Roberto, & Lee, 2002, p. 394).

In their original article, Kotler and Zaltman made the claim that the art of selling cigarettes, soap, or steel may have some bearing on the art of selling social causes (Kotler & Zaltman, 1971, p. 3). Although the lion's share of social marketing has so far been used for personal health (arguably closer to the subject of conventional marketing like consumption from an individual utility and pleasure perspective), campaigns have also concerned recycling, littering, as well as reduced consumption (Peattie & Peattie, 2009).

For example, the following social marketing project aims to reduce consumption with the slogan of 'four P's' in marketing: Product to satisfy the consumer's need; Price, right for the consumer; Place for the product to be available; Promotion of the product by informing about its availability, price, and place. According to Peattie and Peattie, these have been translated by several projects into:

- Propositions instead of products (e.g., you should recycle, or more stuff will not make you happy);
- Accessibility instead of place (and instead of car transport directions);
- Cost of involvement instead of price;
- Social communication instead of promotion (Peattie & Peattie, 2009, pp. 263–264).

In this book's terms, social marketing has mainly been geared towards Apollonian simple trust, supplemented with social communication. Yet this communication seems to be largely about simple transparency, not the wider sub-political involvement of the citizen-consumers. Thus, the reflective Apollonian trust does not seem to be achieved, at least not in the previous example. The marketing is highly top-down-oriented and has thus not attained reflective Apollonian trust. For a sociologist concerned with power struggles, and with the problem of portraying the citizen-consumer as the main source of environmental problems, leaving out government and private firms, this simple Apollonian trust orientation is problematic. Moreover, only few marketing efforts

DOI: 10.1057/9781137276803

have been tried towards a development of Dionysian trust among citizen-consumers in reducing negative environmental impact. Policy practitioners, NGOs, and firms could do much more to stimulate citizen-consumers to be influenced by people who have been successful in their reduction of negative environmental impact.

3.4.3 Group-based interventions towards Dionysian trust

An increasing number of 'group-based interventions' aim at promoting pro-environmental lifestyle change in local regions around the world, particularly in Europe and the United States. The strategy of working with groups instead of single individuals is to make use of what this book calls the Dionysian dimension, where the collective is crucial. An example of the Dionysian dimension in practice is the Green Streets project. The point of the project is to stimulate energy-efficient practice. In addition to attempts at developing an Apollonian trust in the importance and doability of energy saving, its group-based approach is also aimed at Dionysian trust. The project involves peer pressure as well as mutual support in order to make energy-efficient practices stick. In this process, former strangers among neighbours were brought together. Furthermore, the project led people outside the neighbourhood to become interested and involved in energy saving. As Lockwood and Platt state,

> There is clearly some potential here for a virtuous circle between households acting together to reduce energy demand and the strengthening of social cohesion. (Lockwood & Platt, 2009, p. 33)

Moreover, they mention another element closely tied to social values, namely competition:

> A community approach could also allow the wider use of competitions. The competition element in Green Streets was fun, motivating and helped sustain interest. The competition involved two parts – comparison and an incentive (prize). Community prizes (for example funding for local projects or institutions such as swimming pools, football club, church or pub) are likely to be particularly motivating and reinforce the idea of acting together with others. (Lockwood & Platt, 2009, p. 33)

In contrast, a more individualistic design of environmental projects has often been more problematic. For instance, the offering of 'green

DOI: 10.1057/9781137276803

electricity' to households has in many areas involved difficulties in stimulating motivation among citizen-consumers (Ek & Söderholm, 2008, p. 169). One fundamental challenge appears to be that energy use is invisible and 'abstract'. There is little or no way for individual citizen-consumers to expose this green concern to their peers. Also, remember that the ecological values are overridable and typically need to be supported by social values.

From several studies and a comprehensive research review, it is clear that the group-based approach to reducing environmental damage is often successful, in terms of the members completing the project, and in some cases making the practice change stick (Seuring et al., 2008). To mention just one result, in group-oriented EcoTeams in the United Kingdom (UK), 90 per cent of the participants claimed that they had maintained and continued to increase their lifestyle changes to reduce their negative environmental impact (Nye & Burgess, 2008). In a research review of four group-oriented programmes in the UK and the Netherlands involving 1,363 total participants, Fisher and Irvine conclude that

> Overall, where information is available, it seems that the group-based interventions considered here promote lasting change, and that group-based interventions have the potential both to encourage further pro environmental behaviour change in individuals and to inspire them to encourage others to adopt changes. (Fisher & Irvine, 2010, p. 55)

Yet it is important to mention that policy efforts to stimulate citizen-consumers to become environmentally engaged need not only be designed as social-psychological drives to make citizen-consumers 'understand' the environmental urgency (through top-down teaching). It can also be the potential economic benefits of reducing energy use, for instance, that prompts citizen-consumers to act accordingly. The thesis of this book is that such perspectives run the risk of leaving out a wider societal and political debate about new and less vertical, top-down, interaction between civil society, the market realm, and government in environmental issues. To be sure, the preceding design may be 'successful' in a narrow sense and is based on *simple* Apollonian trust, but in this case it is combined with Dionysian trust that efforts be socially beneficial. However, one may also envision 'projects' which are more geared towards a reflective Apollonian trust in combination with the Dionysian trust. Through wider reflective Apollonian trust in the process of acquiring

DOI: 10.1057/9781137276803

knowledge about the ties between many different aspects of citizen-consumer practices, along with the Dionysian trust, citizen-consumer's engagements as groups may have a much more extensive political potential. The engagements may, for instance, involve more discursive political activities such as protests against companies and the government, which are activities that go far beyond narrow household practices in the local environment. Because of information technology, communities can also quickly become global and spread incredibly rapidly. In my view, while overstressing the individuality of political consumerism by defining it as 'individualistic, collective action', Micheletti (2003) provides several examples of how discursive political consumerism may spread far beyond the private household, neighbourhood, or workplace. The collective dimension of political consumerism, as a more social activity than the individual consumers making environmental and ethical choices in a store (sometimes called monetary political consumerism), has so far been much under-researched. The risk is that political consumerism, as practice and research object, remains an example of individual action, with the focus merely on the (overridable) ecological and material motivations, leaving out an integration of the overriding social motivation that is so powerful. As Newholm and Shaw maintain, too much emphasis is placed on the individual and the collective spaces for ethical consumption need further attention (Newholm & Shaw, 2007, pp. 258–259).

3.5 Conclusions and discussion

> 'We try to solve the problems of the world by adopting the Apollonian model only' (Kourvetaris, 1997, p. 234).

Citizen-consumers' typically limited engagement in environmental issues – through monetary, use-oriented, and/or discursive practices – are often explained as (economic and practical) egoism. This explanation can be translated into emphasizing the individual economic goal as the fundamental one. Moreover, this explanation can be modified into the environmental goal perspective if it implies that society – or individuals – may change their economic and practical egoism into an overriding care for the external environment, a concern that may possibly become global in scope. According to these views, society's failure so far has been two-fold. One fault is that society is failing to make reductions in

DOI: 10.1057/9781137276803

citizen-consumers' environmental damage pay; if it paid for all of us to reduce the harm we cause, we would be motivated to make the reductions and to support more comprehensive environmental policies. The other failure argued is scientific/pedagogical: Society has not succeeded in providing sufficient evidence of the urgency of environmental problems. Or, society has failed to package the 'overwhelming evidence' in an optimal way to citizen-consumers for them to make environmental values their overriding ones and act accordingly. To some, these failures may be seen as combinable and by others as incommensurable. For instance, in Krause' s intriguing comparison between societal efforts to make people quit smoking and to stimulate lifestyle change to reduce their contribution to climate change, knowledge-related and information failures are the only ones mentioned; the lack of *Apollonian* trust is presented as central. The problem is described as cognitive; the solution is implied to further strengthen the scientific evidence and to distribute it to the citizen-consumer properly, in turn strengthening her intrinsic environmental values and motivating fundamental engagement.

Compared with smoking, communicating the 'truth' about unsustainable consumption is challenging. Information regarding the consequences of smoking can be clearly presented by deaths per capita and a long list of increased disease rates. However, the 'truth' about unsustainable consumption as a whole is more nebulous (Krause, 2009, p. 288).

Her conclusion is that the citizen-consumer's greenhouse gas emissions should be 'concretized' as follows:

> As was seen with the discussion of cigarette warning labels, *how* information is presented influences its impact. Thus, for example, illustrations converting estimates of average annual energy consumption for each housing option to tons of greenhouse gas (GHG) and ambient emissions, multiplied over a 20-year period, can be expected to have a larger effect on a final choice than a technical footnote. The legitimacy of information is essential, and standardization and government oversight would be important in this regard. (Krause, 2009, p. 288)

As discussed, yielding to Apollonian trust – simple or reflective – has not turned out to be more than partially successful in the case of smoking; so why should it be in the case of climate gas emissions, where the hope for drastically reduced personal health risks and direct money savings is embedded in avoiding smoking, but not in reducing our climate gas emissions? What is missing here, which Figure 3.1 tries to illustrate, is the more fundamental aspect of Dionysian trust. To be sure, learning the

DOI: 10.1057/9781137276803

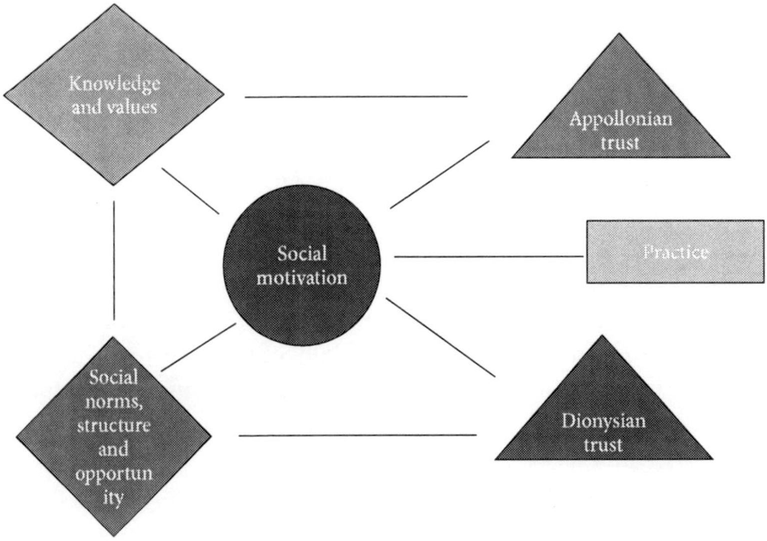

FIGURE 3.1 *Social motivation, Dionysian and Apollonian trust*

enormous figure of our climate gas emissions in tons over a twenty-year period may stimulate reflection about the problems caused by a modern Western way of life, particularly if placed in relation to the average emissions of a world citizen or of an average citizen in the developing world. The Apollonian trust in the knowledge authorities providing data may be very high, and the uncertainties involved in these knowledge claims may be low – even when they are often so dear to us environmental social scientists. Still, in order to go beyond discussions among citizen-consumers who want to ensure social status by showing off such knowledge and by expressing concern among their peers, comparisons within one's own social groups and across groups with similar socio-economic categories are much more likely to trigger social interactions, Dionysian trust in forerunners, and so forth, that in turn may lead to changes in norms and practice. This is articulated in a convincing way by Lockwood and Platt, whose case study is presented in the following. They hold that,

> Rather than a crude and simple comparison of energy use with a national average, smart metering could allow the averaging and comparison of data across streets, neighbourhoods, non-geographically defined communities, cities etc., with prizes at different levels. (Lockwood & Platt, 2009, pp. 33–34)

DOI: 10.1057/9781137276803

This idea is based on the acknowledgment of the more motivating Dionysian trust, based on feelings of wanting to do well in the social game, feelings developed millions of years ago, and much in line with what the classical sociologists referred to as a key component of *Gemeinschaft*, community, primary groups, mechanical solidarity, and so forth. The connection between these classical sociological concepts and the Greek dichotomy is further supported by the fact that these scholars were all directly influenced by Apollonian and Dionysian ideal types (Kourvetaris, 1997, p. 233).

Nevertheless, it deserves to be repeated that Dionysian trust is of little value for reduced environmental harm over time if this trust is not accompanied by an Apollonian trust, ideally of the reflective kind. Even Nietzsche, who may be the least diplomatic among our major thinkers, and who considered himself 'a disciple of Dionysus', maintained that society and arts should struggle towards a balance between Apollonian and Dionysian traits, in our lives as in the arts (Nietzsche, 1872). Not least in present-day society, with extensive disembeddedness between our daily routines and their environmental as well as social consequences, and with our dependency on expert systems providing us with knowledge about such ties, the necessity of Apollonian trust is obvious. Nevertheless, to stimulate long-term green political engagement of citizen-consumers, social institutions need to create bases for this trust to be of the *reflective* kind. This presupposes transparency and dialogue about the environmental claims, their foundations, and the participants in this process of knowledge acquisition. Moreover – and this is the most overlooked aspect of environmental problems, information, and trust: the *Dionysian trust* in people proposing or demonstrating practices of reduced environmental harm. Counter to being 'irrational', however, as is often claimed when Dionysian traits are brought up, Dionysian trust should be seen as a trust in appeals to our social motivation: the assurance being influenced by other people's practices, such as reduced use of energy and car use which may be beneficial for our social position, not least among our peers, friends, colleagues, neighbours. To be sure, positive environmental consequences – from local to global – may be extensive, but still remain *means* or *side effects*. That is how we are built.

Notes

1 The *Gemeinshaft–Gesellschaft* distinction has similarities to Durkheim's distinction between mechanic and organic solidarity, but a discussion of

similarities between these two dichotomies is beyond the scope of this book.

2 As Magnus Boström and myself discuss in another book, there are two more layers of transparency that mainly concern Apollonian trust: intra-frame and inter-frame transparency. The former refers to visibility about the extent to which criteria and procedures are consistent with the dominant problem frame (a frame, such as climate change, chemical risks, biodiversity, nuclear power risks), and inter-frame transparency, referring to the extent to which 'the dominant frame should be altered to include more important goals' (Klintman & Boström, 2008, p. 180). It is beyond the scope of this chapter to go in depth with these layers.

DOI: 10.1057/9781137276803

4
Rebound Effects and Spillovers

Abstract: *Relations between a citizen-consumers' practices are examined. To be sure, citizen-consumers may be ecologically inconsistent. Yet, a citizen-consumer buying second-hand clothes and flying to exotic places can be 'lifestyle-consistent'. Rebound effect and spillovers are analysed as risks and potentials of energy efficiency, and savings for citizen-consumers' other environmental impact.*

Klintman, Mikael. *Citizen-Consumers and Evolution: Reducing Environmental Harm through Our Social Motivation*. Basingstoke: Palgrave Macmillan, 2013.
DOI: 10.1057/9781137276803.

 DOI: 10.1057/9781137276803

This chapter examines citizen-consumers' inclination, or lack of inclination, to perform practices on the basis of their environmental outcomes on a global scale. By practices, I refer to the parameters of political consumerism, namely monetary (choices of products and services), use-oriented (how citizen-consumers use, or do not use, products and services while 'leading their daily lives'), and discursive (verbal and written activities concerning the environmental and welfare consequences of modern daily life). What are the relationships between what citizen-consumers say and what they do with regard to environmental consequences? What can be asserted about so-called 'inconsistencies' across the different parts of our lifestyles concerning degrees of environmental considerations and outcomes?

As we shall see, the ecological motivation approach typically perceives preferences as a tension between materialist and non-materialist values. Yet, according to the ecological motivation approach, preferences can be affected by political and social change so that ecological values become overriding. The material motivation approach, however, treats preferences as fixed, oriented towards material utility, comfort, and so forth. Individuals are highly motivated to modify their choices so that these individual ends can be reached through minimal material/economic effort. In a society where many people find environmental damage problematic, the position of the material driver contends that the main solution to environmental harm is to create policy arrangements through which reduced environmental harm is financially beneficial to the citizen-consumer, whereas increased harm is costly. From this book's position, however, ecological values are important, but they can never be intrinsic. As to the material dimension, it is certainly crucial in making environmentally sound solutions pay and harmful ones cost. Yet this is predominantly important when firms and companies are concerned, in particular those organizations which have economic profit as the main or only goal. As to citizen-consumers, their preferences are different.

4.1 Doing, saying, and believing as social practice

This section gives three interpretations of the relationship between what citizen-consumers do, say, and believe concerning the environmental implications of their lifestyles.

DOI: 10.1057/9781137276803

4.1.1 Ecological hypocrisy

According to scholars and practitioners that emphasize the importance of primary ecological values (either biologically inherent or culturally developed), modern society has long entailed a distance between nature and society. In the modern world, this distance is termed by some social scientists as (social) disembeddedness, but it leaves a gap, an invisibility, between citizen-consumers' practices and the environmental as well as wider social consequences of these practices. The disembeddedness can be seen in people's limited environmental efforts and in environmental attitudes to the extent that attitudes indicate low levels of concern for the environment. Yet, according to the ecological motivation approach, the high levels of environmental concern that can still be seen through environmental attitudes in certain parts of the world may be a sign that changes are taking place towards a higher degree of environmental reflexivity and concern (Beck, 1999; Giddens, 1991; Inglehart & Abramson, 1999). Also, policy measures towards social re-embeddedness of the relation between citizen-consumers and the natural environment – measures such as eco-label, standards, environmental information programmes – have contributed, along with environmental media attention, to many citizen-consumers' trust in these environmental information sources as corresponding with and strengthening people's primary ecological values. They have also served as a basis for questioning mass consumer society. Nevertheless, the ecological motivation approach acknowledges that there are at the same time heavy social structures such as (sub-) urban planning, automobiles causing risks to bikers, pressure from tenant owners' associations, globalization of production, and so forth that work against people's current or potential wishes to reduce environmental harm (Schor, 2008). According to this approach, all of this constitutes possible gaps between people's (often strong) environmental attitudes and (still often low) environmental practice.

4.1.2 Constant material preferences

How does the ecological motivation approach compare with a material motivation approach? According to the latter, citizen-consumers are basically materially goal-rational individuals (by genetic evolution or by culture) leading their lives by trying to make economically sound choices to lead a fulfilling life. To be sure, the material motivation approach does not need to go to the extreme of portraying a *Homo economicus* in a

DOI: 10.1057/9781137276803

schematic, stereotypical way (Wilson & Dixon, 2008). Citizen-consumers may be affected by risk society's late modern calls for environmental concern. Also, several studies in economics indicate citizen-consumers' willingness to pay for environmentally superior products and services. Yet, these studies, when focusing on a representative sample of an entire population, often indicate either citizen-consumers *stating* a limited willingness to pay environmental price premiums or citizen-consumers' even more limited actual payment of environmental premiums (Grolleau, Ibanez, & Mzoughi, 2009).

Scholars and policy practitioners endorsing the material motivation approach admit that material incentives and disincentives run the strong risk of losing their strength as soon as these policy instruments are removed. This temporariness is logical if one assumes that the material values override the social motivation and ecological values. Support for the view of temporariness is shared with scientific studies pointing towards drastically reduced consumer support such as for 'eco-friendly automobiles'; support declines as soon as subsidies are removed, unless there are public expectations that fossil fuel will become more expensive (Turaga, Howarth, & Borsuk, 2010). To be sure, ecological market segments have grown very rapidly in many sectors, particularly the demand for organic food. Yet this position contends they will remain niches as long as price premiums are high.

4.1.3 Socially consistent practices

According to a social motivation approach, it is fair to say that studies of citizen-consumers have typically shed light on 'inconsistencies' between what people say and what people do. However, this distinction is typically based much more on the *researchers'* biased focus on issues that are not necessarily at the heart of the citizen-consumers.

From an environmental point of view, to understand what may be perceived as gaps between what people say and what they do, we are helped if we instead see as *practices* both what people say and what they do. According to the social motivation approach, to say that we care about the environment, not least on a global scale, is understood as an *act* that signals (to ourselves and to others) that we are both intelligent and compassionate, two factors crucial for how we act in our social relations since the dawn of humanity. However, it is also important to note that we do not have to be lying when we express concern without having 'physical' practice follow. How deeply our environmental concern actually runs

DOI: 10.1057/9781137276803

(if that can ever be known) may not matter, because human nature has incorporated self-deception as a crucial factor for citizen-consumers to be able to convince others in the social game that they are decent, dependable persons.

Moreover, to admit inconsistencies between what we say and how we actually lead our lives may be more socially valuable than the two alternatives, namely (a) to pretend that we do more for the environment than we actually do, or (b) to pretend that we do not care as much for the environment as we actually feel that we do. By signalling our inconsistencies, we add the trait of honesty to the duo of intelligence and compassion, thus constituting three highly beneficial factors in the social game.

What is crucial here is the *effect* of our signalling, rather than how deeply our actual environmental concern runs, and how honest it is. The social effect of our signals and verbal claims of environmental concerns is what matters: that other people and groups perceive this person as intelligent, compassionate, and honest. In fact, this effect may have strong social value. As to late modern ambivalence, so fondly stressed by many social scientists (Bauman, 1993; Cohen, 2010; Ricci, Bellaby, & Flynn, 2010) and myself included (in Boström & Klintman, 2009), the ambivalent outlook, or mindset, is certainly real and very intensive among many citizen-consumers; for instance, regarding how to lead their lives in ways that harm the environment as little as possible. Still, one should ask whether ambivalence is only a useless frustration or to what extent it may play a productive role in the social game. From this perspective, expressions and discussions about one's ambivalence can be seen as an art: of going back and forth with arguments for and against, combining and weighing different types of knowledge, and so forth. Moreover, ambivalence resonates with Socrates' famous assertion that wisdom means acknowledging how little one knows for certain. The basis of ambivalence is *arguing*, which has been crucial to mankind ever since the ancestral environment – crucial both for solving substantive problems (how to improve hunting and gathering), and in terms of procedure, thus strengthening our position in the social game. Yet, in modern society, when we as citizen-consumers debate the environmental challenges in our daily lives, the ambivalence does not end; and according to the principle of reflective modernity, ambivalence *should* never end. The moment our sense of ambivalence ends we might be perceived as naïve, static, and stuck in simple Apollonian trust, which is not the way to lead a reflexive modern life. Consequently, by exposing and elaborating on our gaps between environmental concern and practice, we might – at least until norms change thoroughly – be seen as more intellectually advanced. This inconsistency between attitude and practice may leave us better off in our social relations than a strong environmental concern combined with full, simple trust in society's environmental information and prescriptions.

4.2 The relationship between practices

To highlight how the three positions analysed in this book perceive citizen-consumers' lifestyle choices, it is illustrative to go beyond a discussion about the various meanings of *one* specific choice, to the relationship *between* different choices in daily life.

4.2.1 Ecological consistency through knowledge

According to the ecological motivation approach and in terms of environmental impact, the relation between the various parts of a citizen-consumer's lifestyle may become an intrinsic basis for that person's feeling of contentment. For instance, if there is much inconsistency between a citizen-consumer's trips to work compared to her leisure travel, this may create much cognitive dissonance and frustration, ultimately making her modify the less ecologically sound travel modes in order to become a consistent environmentally friendly citizen-consumer. All practices have environmental impact, but knowledge authorities may help citizen-consumers acknowledge particularly urgent problems, and what changes in practice are needed. In modern society, there is serious inconsistency between different parts of our lifestyles. It may also be that people are unaware of the environmental impacts of some parts of their lifestyles, an ignorance that contributes to this ecological inconsistency. A main solution, aside from hard regulation and heavy taxation on, for example, certain modes of travel where more environmentally beneficial modes are available, is extensive and innovative modes of environmental information. This may help people become more ecologically consistent (ideally in a positive way) by providing information about ecological footprints. Life cycle analyses (LCAs) on products and services would make environmentally 'invisible' factors and impact visible because we can compare various alternative products and services. According to the

DOI: 10.1057/9781137276803

ecological motivation approach, taking such calculations into account and comparing their lifestyle parts may in itself be a potent source of lifestyle change for citizen-consumers. Researchers in social psychology with a particular interest in 'environmental behaviour' have made ample efforts to find a degree of consistency across parts of citizen-consumers' lifestyles. Several theories back up the hypothesis that environmentally beneficial practices tend to 'spill over' to other domains in our daily lives (Frey, 2000). Many social psychological theories emphasize the need for humans to avoid or lessen inconsistencies between attitudes and behaviour (Cognitive Dissonance Theory and Balance Theory; Eagly & Chaiken, 1993). The applications of such theories to environmental issues attempt to shed light on the human struggle to reduce *environmental* inconsistencies. As the environmental social psychologists Thøgersen and Crompton state,

> If this is true, these people would more often than others, everything else being equal, behave in an environment-friendly way in any specific domain. (Thøgersen & Crompton, 2009, pp. 149–150)

But, according to the ecological motivation approach, what can be spilled over across lifestyle domains triggering an increased environmental consistency? Two factors are particularly in concert with the ecological motivation approach. The most obvious one is for the citizen-consumer to spread the ecological value from one practice to another, making further lifestyle domains ecologically value-ridden. Yet, there is another spread of factors that are fully in line with the ecological motivation approach. A rather common view among researchers is that it is not people's values that are the obstacle to environmental adaptation. According to this view, people in modern society may already have the fundamental ecological values needed. What they lack, and what could be spilled over across lifestyle domains, is the knowledge, skills, and opportunities needed to act out one's ecological values. In my terminology, their approach contends that Apollonian trust in the alteration of practices is missing. In line with that view, the social psychologist De Young contends that environmentally positive spillover may take place when people acquire 'procedural knowledge' from one practice, knowledge that can be applied to another lifestyle domain (De Young, 2000). A preference for eco-labelled fish (MSC) is an area that Thøgersen and colleagues have found to correlate with the likelihood of choosing other eco-labelled goods. Thøgersen explains this with the gradual routine and knowledge that the active choice of one type

DOI: 10.1057/9781137276803

of eco-labelled products may generate. Still, he admits certain limits to spillovers of eco-practice through the spillovers of knowledge:

> One may speculate that the likelihood of spillover due to such learning processes might decrease rapidly with increasing dissimilarity of behaviours. (Thøgersen & Crompton, 2009, p. 149)

4.2.2 Constancy of material use

According to the material motivation approach, the relation between a citizen-consumer's various practices may well be inconsistent from an environmental perspective. Yet understood in material terms, her practices are highly likely to be consistent. According to this position, because the material motivation is more fundamental than ecological values, a sense of cognitive dissonance due to 'environmental inconsistencies' is not likely to run very deeply concerning her own lifestyle. On the other hand, political frustration may emerge in her concerning the lack of political measures to ensure that environmentally sound choices pay, and that unsound ones cost more or are banned.

This approach contends that the citizen-consumer typically selects products and services that give her good value for her money. To the extent that eco-innovations have gone from niches to mainstream in various sectors, she is likely to choose products that are highly 'environmentally progressive' in certain sectors or parts of sectors, such as work transport, but not in others, such as leisure travel because the latter is still done largely by aeroplane or car. Her preferences are most often based on convenience and pleasure: it is thus obvious that citizen-consumers from the North want to go to warmer places for holidays or to beautiful ski resorts, as intrinsic preferences. The common choice of car use instead of cycling or public transport is also most basically founded on convenience, practicality, comfort, time saving, and so forth. Moreover, from an environmental perspective, because the material motivation approach conceives money and practicality as the bases for choices in our lifestyles, this position is likely to entail an acknowledgement of the risk of negative rebound effects.

Despite the importance that the material motivation approach ascribes to making environmentally sound solutions pay to attract citizen-consumers, scholars as well as some practitioners of this position have been particularly well suited to acknowledge a risk here. To the extent that citizen-consumers are driven by the interest in getting maximum

DOI: 10.1057/9781137276803

material value from their money, what might this entail on a market that, in the name of environmental sustainability, *elevates increased energy efficiency* of products and services to the ideal? Environmentally negative rebound effects can be such a risk. Rebound effect ('take-back') refers in this context to the situation where reduced environmental harm from one practice may lead to increased *negative* environmental impact in another area of practice. Renowned economists have highlighted the rebound phenomenon, even aside from environmental impacts, from Jevons in 1865 to present-day researchers. (For comprehensive overviews, see Hertwich, 2005; Steven Sorrell, 2009). To be sure, there is much scientific disagreement about the extent and levels of the rebound effect. Some scholars estimate the effect to be from low to moderate (Greening, Greene, & Difiglio, 2000), whereas others referred to in this section estimate a high level, or different levels among different socio-economic groups and gender (Carlsson-Kanyama & Lindén, 2007; Wadud, Graham, & Noland, 2009). However, it is fair to say that very few deny the existence and a certain degree of importance of the rebound effect in current lifestyles.

In the following, I briefly describe the rebound effect through the lens of the material motivation approach. When a citizen-consumer purchases, for instance, a fuel-efficient car – mainly for economic reasons with supposedly environmentally positive side effects – money is supposedly saved. Given people's fundamental interest in maximizing material value, which includes comfort and convenience (in sum fairly straightforward hedonism), the money saved is much more likely to be used on something else that is environmentally harmful. This rebound may often outweigh the environmental gain of the energy efficiency. The rebound effect is *direct* when the saved money is used for more of the same practice, in this case longer or more frequent trips. The effect is *indirect* when the saved money is spent on another practice, in the case where it harms the environment. This other practice could be tourist travel to a remote destination, but the money could also be spent on a motorcycle, more home electronics, or any other goods or services with negative environmental impact. The principle behind the negative rebound effect is that citizen-consumers have a certain 'fixed' amount of resources, particularly money and time. Citizen-consumers have a primary driving force geared to making use of these resources as 'rationally' as possible, in terms of getting maximum value for the money and time. This value is typically materially related to comfort,

DOI: 10.1057/9781137276803

convenience, and pleasure, the latter in strong correlation with modern consumerist society. Therefore, energy-efficient vehicles, rather than reducing overall pollution in society, may entail more people moving further out into the suburbs, to more spacious and comfortable houses where the land prices are lower. A more or less necessary outcome of energy efficiency is longer car trips, and larger spaces to heat up and cool down.

4.2.3 Response from the social motivation approach

Like the other positions, the social motivation approach acknowledges the possibility of lifestyle inconsistency from an environmental perspective. However, in principle, the social motivation approach is closer to the material motivation approach in assuming that environmental consistency is not an essential inconsistency among citizen-consumers. As to the social motivation approach, our primary values concern social relations, in addition to basic material needs and the motivation towards creativity, invention, and skills. From this position, one may assume that there are struggles, conscious or subconscious ones, not so much to achieve social consistency as to improve how one does in the social game. What is consistent is that this social struggle is one where citizen-consumers may be more successful in some parts of their lifestyles than in others. Importantly, efforts that concern maintaining or strengthening their position in social groups may overlap environmental efforts in some lifestyle areas but not in others.

What can be said about environmental spillovers and rebound effects from a social motivation approach? As to environmental spillovers, they may well occur and be interpreted from a social motivation approach. However, where spillovers occur are not because of a fundamental need to be environmentally consistent. Instead, they are due to the motivation to be consistent socially and often 'aesthetically'. However, this motivation is not a simple and easily interpreted one. As citizen-consumers, most of us want a certain tension somewhere between aesthetic consistency, group identity, and individual originality. What the cognitive, social psychological approaches often miss is the importance and social motivations of this intricate tension rather than of full environmental consistency – or consistency in any other moral domain. If there is a struggle to achieve consistency among citizen-consumers, it is more likely a struggle for consistency in a social, aesthetic sense, with a motivation to do well in

DOI: 10.1057/9781137276803

the social game across lifestyle activities. Nevertheless, environmentally beneficial spillovers may take place and become significant in this socially oriented struggle, albeit as a secondary value, as a means or a positive side effect towards the meeting of primary social motivation. Thus, spillovers may well take place and become powerful from purchases of eco-labelled products in one domain to another domain, or if a change from car use to bicycle use leads to political engagement in green urban planning.

This is in line with several empirical studies of travel and lifestyles. For instance, the geographer Barr and colleagues have examined relationships between people's tourist practices and their everyday lives, focusing the study on environmental aspects. The title of these scholars' article is very telling: 'A Holiday is a Holiday'. They found a large gap between these two sets of practices in terms of environmental consideration and in environmental outcome, which has been confirmed in other studies. Whereas many people do quite a lot in their homes in the name of the environment, tourism is typically viewed differently, as less environmentally important, despite its substantial environmental impact (Barr et al., 2010). The tourism scholar Hares and colleagues have drawn the same conclusions. They have studied UK tourists' views and travel decisions, and have concluded that people in general, who consider the environmental consequences of daily practices, are very likely to leave this dimension at home when they are at the travel agency planning for their flight trips. Thus, people segment different parts of their lives, putting considerable emphasis on the environmental implications of one part while more or less ignoring such implications in another area. My own empirical studies of household practices indicate a similar pattern, not just concerning verbal preferences across various practices, but also in terms of the relative environmental impacts across practices.[1] Barr and colleagues, who were previously mentioned, explain this segmentation with reference to low-cost air travel, fully in line with the material motivation approach. This also corresponds to the challenge of the rebound effect: If people have saved money on energy-efficient household practices, the money is used for something else. If the cost of air travel were several times higher, on the other hand, fewer people could afford it. Still, the deeper roots of the problem of the rebound effect, and the potential for environmental spillovers, is not that the middle and working classes have too much money or that people have not been indoctrinated enough to think environmentally. Rather, the root of this issue is that practices associated with reduced environmental impact so far have not been designed or developed to take into account the primary value of social relations, values

DOI: 10.1057/9781137276803

more deeply rooted than environmental concern or material accumulation. In other words, both phenomena are certainly possible, but they are by no means obvious, nature-given, or hopeless to influence. If policy makers and companies collaborated with groups of citizen-consumers in meeting the deeply rooted social motivation in environmental projects, levels of rebound effect and spillovers would likely be affected. Alternative types of travel and travel modes could be developed in which negative environmental impact is reduced and social bonding is strengthened.

In contrast to the ecological and material motivation approaches, the social motivation approach indicates the potential for changes in daily practices without or with very modest inclusion of material or ecological benefits. Riding one's bicycle to work, for instance, may under some circumstances be beneficial for how one does in the social game: it signals health, balance (beyond not falling off the bike but in one's life more broadly), environmental concern, and self-assurance in not having to show off an expensive car. This is termed post-materialism (Inglehart & Abramson, 1999) or 'conspicuous simplicity' (Ballantine & Creery, 2010). Yet, flight trips to areas of distant 'eco-tourism', despite their many tons of environmentally harmful emissions, may very well be 'aesthetically' consistent with the practice of cycling to work daily. Moreover, they may not only be consistent, but the actors may also be interdependent. This combination signals a person environmentally concerned in word and deed, which indicates knowledge and compassion. Moreover, the person may manage her time well, despite having an agenda full of activities. Importantly, her bicycling is not based on poverty. It is her free choice, even if it may not be the most time-efficient mode of transport in all situations. Her other choice, of eco-trips to exotic places in her leisure time, indicates that she has solid financial resources, so solid that she may afford the premium costs for the ecological profile of the trips. Although one's material position does not overlap social position in any full sense, and although material accumulation is an overridable value, having a solid financial situation is typically something that, when exposed in a way that is not perceived as vulgar, is beneficial for how one does in the social game. The social psychologists Thøgersen and Crompton, although elsewhere suggesting the subscription to an ecological motivation approach, admit the importance of social status for adopting 'environmental behaviour', although these authors make the mistake of presenting a wide range of factors as being just as fundamental as the social ones (Thøgersen & Crompton, 2009).

DOI: 10.1057/9781137276803

4.3 Conclusions and discussion

This chapter has examined relations between various things people do, as well as between reasons for doing these things. The general answers, from the social motivation approach, can be summed up as follows:

- *What citizen-consumers say and what they do may seem inconsistent.* It is not rare, for instance, that people are very worried about pollution and climate change, find their car use environmentally problematic, realize that there are alternatives to car use, but they continue – or even increase – their car use. Yet what people say may also be perceived as practice, and not merely something that is – or should be – a perfect reflection of their 'real acts'. What people say is often, subconsciously or consciously, a practice influencing how well people do in the social game. Verbal expression of environmental concern is rarely lying; it is typically honestly felt. Yet it may have traits of self-deception – something that has had great evolutionary value, because it makes us perfectly honest in the eyes of others and at the same time it helps the citizen-consumer to do well in the social game, by signalling knowledge, awareness (intelligence), and compassion.
- *When comparing what citizen-consumers do in various domains of their lives, people's practices may also at first sight seem inconsistent, at least from a perspective of environmental sustainability.* Many of us are highly engaged in purchasing, using, and debating about some daily practices from an environmental perspective but much less so in others. Is this 'inconsistent'? It is from the ecological motivation approach. How can we have the deepest environmental concerns when, for example, we select an eco-friendly fish product in the store, then ignore eco-friendly labels when choosing wine to drink with the fish? The answer using the ecological motivation approach would perhaps be that citizen-consumers are ignorant about the environmental implications of some products, or ambivalent about the reason for holding Apollonian trust in the knowledge validity behind ecological marketing. The answer from the material motivation approach would perhaps be that there are limits to how much premiums we want to pay for ecological products, or that we may simply enjoy the non-ecological wine and, therefore, prefer it. In this specific case, the social motivation approach is fairly close to

the economic one: choices in our daily life are not fundamentally ecological to us. Our deeper motivation, conscious or subconscious, is rather to do well in the social game, in this case by impressing our social groups (or strengthening our identity) with an aesthetic competence, generosity, and so forth (Griskevicius, Tybur, & Van den Bergh, 2010). This lesson is a necessary one from the reading of Darwin, particularly with sociological glasses (Reich, 2010). If our practices also have environmentally benign implications, while at the same time signalling compassion and awareness, that is a positive means or side effect.

- *When examining reasons behind people's lifestyle practices that some perceive as green, the ecological motivation approach obviously emphasizes the importance, or potential, of this motivation, and the power of this motivation in policy making, advertising, and so forth.* On the other hand, the material motivation approach stresses that green choices also must pay economically as well, or that they should not cost too much if they are to make people motivated. The social motivation approach, however, may well admit that ecological and economic factors are indeed powerful, particularly in combination. Still, the power of such a combination depends on the extent to which these factors are supported by the norms and culture of the groups we belong to or wish to belong to. Moreover, several social aspects go deeper than the factors of environment and economy. A policy implication of this is that people's readiness to purchase, use, or in other ways engage in environmentally sound products or services is not necessarily dependent on environmental reasoning. Nor are such practices necessarily determined by substantial material compensation. For example, mobility management (MM) programmes that stimulate people to use a car less and instead use more environmentally friendly modes of transport have shown how people who adopt the changes prescribed by MM often do this for several reasons other than environmental benefits. One's own health is one important factor, both inherently (meeting basic 'material' needs) and as a factor that may converge with the social driving force. The curiosity about a new 'lifestyle project', requiring new skills, knowledge, and routines, was also a motivating factor mentioned in interviews with MM adopters. Intriguingly, the environment was hardly mentioned by the MM adopters in the open-ended interviews.

DOI: 10.1057/9781137276803

Quantitative attitude studies, surveys, and market investigations with a focus on environmental attitudes and behaviours are often used as the empirical basis for research on citizen-consumers and environmental concern. Although some of them are conducted with high scientific rigour and precision, common mistakes in such studies are that they presuppose ecological or material accumulation-oriented values – or a combination of these two – as actually or potentially constituting *primary* values of citizen-consumers. This is typically reflected in questions such as 'How much – if anything – are you willing to pay extra for sustainable products?' or 'How much extra travel time – if any – would you accept for choosing train over car?' Citizen-consumers are here seen as isolated individuals driven by a trade-off of ecological and material motivations only. Interestingly, similarly designed questions are also common within the framework of political/ethical/concerned consumerism, a theoretical perspective which is perhaps expected to embed an awareness of social dimensions of consumption and lifestyle choices.

- *The social motivation approach underlines the importance of how we as citizen-consumers think that people around us in the groups we belong to or wish to belong to perceive various lifestyle choices.* This does not mean that citizen-consumers always try to follow the established norms of these groups. Nevertheless, it means that people find it important – consciously or subconsciously – to relate to the norms and practices of these groups, either by conforming to or diverging from them, the latter alternative possibly initiating minor or major changes of these groups and practices. The social motivation approach can appear as cynical. Yet it must be remembered that conscious struggle for social status is only one of the multiple dimensions of social relations. The part of this struggle that is manifested in competing with others is an even narrower trait within social relations. Friendly peer pressure, from family, friends, colleagues, and other people on the street, is at least as common and important a part of social relations. It remains a crucial challenge of how this aspect of social relations may be strengthened in policies developed in dialogue with groups of citizen-consumers themselves.

DOI: 10.1057/9781137276803

Note

1 In my studies, conducted in Sweden, the UK, and the Netherlands, it was clear that people with less of an environmental rhetoric sometimes performed at least as many and frequent practices associated with environmental concern: recycling, composting, bicycling, albeit with explicitly non-environmental rationales (Klintman, 2000). Such rationales could be that the neigbourhood had agreed on it, that it was healthy (biking), and so forth.

DOI: 10.1057/9781137276803

5

Single Policy and Planning Issues

Abstract: *Three factors are analysed: (1) citizen-consumers' acceptance of green policy and planning that goes against short-term material benefits; (2) citizen-consumers' participation in 'green' planning; and (3) citizen-consumer environmental activism. Initial citizen-consumer scepticism in 1 to 3 may change into support and engagement. The chapter shows how social motivation can facilitate such change.*

Klintman, Mikael. *Citizen-Consumers and Evolution: Reducing Environmental Harm through Our Social Motivation*. Basingstoke: Palgrave Macmillan, 2013.
DOI: 10.1057/9781137276803.

 DOI: 10.1057/9781137276803

As mentioned in Chapter 2, the term 'citizen-consumers' refers to a main hybrid role that we have in society. With regard to environmental issues, this implies that there are several ways in which we could have positive or negative impact. Whereas previous chapters have to a large extent been preoccupied with the consumer role, particularly monetary and use-oriented consumerism, this chapter seeks to widen the perspective. This is done in three ways, namely by discussing (1) public acceptance of environmentally progressive policy and planning; (2) public participation in various schemes marketed as 'green'; and (3) more activist ways in which citizen-consumers may be involved in environmental issues, by protesting against companies and products, but also against politicians and policy makers, if these actors and institutions show a weak environmental record.

5.1 Progressive, environmental policy making

Although the environmental impact of our daily purchases and uses of products and services is often highlighted, some scholars contend that acceptance and subscription of progressive environmental policy making and planning is an area where citizen-consumers may have the highest environmentally positive impact. After all, politicians in democratic societies are dependent on public approval of their work in order to be re-elected. Based on this truism, it is easy to conclude that dramatic improvements of society's environmental record could be made rapidly, *if only* citizen-consumers showed that they accept the governmental actions needed. What – if anything – can trigger us to accept progressive environmental policy making that might hamper our short-term individual freedom, while benefitting society as a whole and its members in the long term?

As earlier chapters illustrate, advocates of the ecological motivation approach would assume that such public acceptance is possible, provided that the environmental facts are presented sufficiently and understandably to the general public. These advocates would believe in the strength of environmental facts, and their potential for being directly tied to basic and practice-triggering values, such as support for comprehensive policy changes like raised eco-taxes and so forth.

Moving to the issue of policy acceptance from a material motivation approach, this approach would imply a contention that citizen-consumers'

DOI: 10.1057/9781137276803

policy preferences are fairly fixed. Because the bottom line of this approach is that citizen-consumers generally prefer to get the highest number of quality goods and services for their money, quality in the sense of high direct material and perhaps aesthetic value for the individual, policy preferences, as well as the effects of a certain policy measure, ought to be more or less constant.

Although not necessarily explicit, this latter approach is reflected in a wide array of research and policy analyses that base their conclusions on surveys conducted at a single point of time. This approach implies that such surveys may give a good idea of people's attitudes, because preferences are fairly constant as well as conscious to people ('Everyone knows that money matters'). If people find out how much the new policy will cost them, in money, time, and perhaps also the environmental benefits, they should get a good grasp of the extent to which they are willing to accept the future policy change. If the change is costly, people are unlikely to accept it, despite potential environmental gains. If the policy is nevertheless implemented, people will weigh time and money, and above a certain level they will choose alternative practices or protest against the new policy.

Concerning public acceptance of increased taxation and fees on environmentally harmful emissions, some studies draw conclusions implying what public preferences would consistently be. For instance, some studies indicate a public preference for soft and voluntary measures, such as information on environmental risks (Schumacher, 2010). From the results of a Swedish survey in 2002, Jagers and Hammar conclude that people dislike policy modes that constitute disincentives such as increased CO_2 taxes on petrol. The survey results are consistently interpreted as showing that people as a rule prefer the state to use incentives rather than disincentives in environmental policy incentives and 'soft' measures, such as environmental information and decreased tax on fuels that do not affect the climate. In another study overviewing research, Lundqvist concludes that policy measures that are directed to the individual polluter, as a direct response of individual practice, such as car use, are considered less preferable than a general eco-tax that affects the whole population equally, even if this makes general taxes more expensive for each taxpayer (Lundquist, 2006, p. 25). These conclusions, and the analyses behind them, seem to reflect a mixture of a material motivation approach, yet accompanied by Lundquist's interpretation of a social motivation that overrides economic irrationality.

What these two studies have in common, however, is a view – at least implicit – that citizen-consumers' preferences are fairly fixed. Moreover, the authors assume that the policy preferences in Sweden, the country where the latter studies have been conducted, can be translated to the preferences in other countries. If the tax-friendly country of Sweden shows scepticism about directed eco-taxes, other countries are likely to show the same pattern, yet located further to the sceptical side of the scale.

How would an evolutionarily based social motivation approach interpret empirical data on public readiness to accept environmental politics and policy measures, data that appear to go against short-term individual interests? First of all, we should ask whether public acceptance in such a direction is at all possible? It is possible. To mention just a few examples, political preferences, although they mainly follow the economic conditions of the individual voter, sometimes go against the economic position of the individual. In the United States, for instance, it is not uncommon that people from the more economically vulnerable part of the working class to support the Republican Party, which is traditionally the party more critical of unions, regulations of a minimum wage, and so forth. Among the middle and upper middle class in several parts of the world, support for fairly high taxes is not uncommon, despite the fact that this might be financially disadvantageous for individuals in this class. Both of these examples indicate the role that the social norms of the respective groups play for political preferences. Still, if social norms are to be treated seriously, the social motivation approach requires that they be analysed, or at least discussed, with a longitudinal perspective over time. As regards *present* social norms about environmental policy measures, the previously mentioned conclusions about people's preferences seem admittedly to be correct; people in several corners of the world seem to prefer environmental carrots to sticks, at least when directed to citizen-consumers. (Environmentally 'evil' companies raise entirely different, far less tolerant sentiments among the general public.) Currently, it is far easier for politicians and policy makers to introduce incentives than disincentives. Yet, this current public preference is no law of nature. It only means that the dominant social norm *of today* has not yet reached a stage where citizen-consumers are in favour of harder environmental policy measures, aside from policy measures to address individuals and groups who affect the environment substantially more negatively than the average person in the Western world leading a suburban, middle class way of life.

DOI: 10.1057/9781137276803

Here it might be refreshing to compare with social norms in other areas of society. Even without the illusion that all things get better over time, one should acknowledge that some things have done so, at least in certain parts of the world. As to the dominant social norm regarding sexual harassment at the workplace or racist expressions in the public realm, a few decades ago a substantial part of the general public in the Western world might have preferred softer measures to handle these issues, such as information and incentives for those abstaining from such practice. As of today, regardless of the popular academic prescription of super-soft 'nudging', suggestions of incentives, gifts, or other benefits for people who do not engage in sexism or racism would probably be seen as both ridiculous and obsolete. Instead, the dominant social norm has in some cultures been changed into one requiring people to abstain from overt racism or sexism, and people who still persist can expect to be punished in one way or another (when it comes to covert or indirect racism and sexism, however, we still have a long way to go). Put in a value-laden language, the progress of social norms in the Western world has been much greater than that of social norms concerning environmental harm.

BOX 5.1 *Changes in public preferences: congestion fees in Stockholm*

In 2006, attitudes towards the idea of introducing congestion fees in Stockholm were measured in the region. A majority of the respondents had negative attitudes to the idea. Nevertheless, a full-scale trial period with congestion fees was initiated in Stockholm the same year. As it turned out, almost half of the respondents changed their attitudes during the trial period, most of them from negative to positive. It should also be mentioned that the attitudes reflected in media were changed from skeptical to positive during the same period. In September 2006, a referendum was carried out about moving from trial period to 'definite' of congestion fees. The majority of citizens in the region voted in favour of congestion fees, entailing a renewed introduction of congestion fees in Stockholm in August 2007. Winslott-Hiselius and her colleagues who have examined the process closely, draw the following conclusions: 'It was not so much that knowledge about environmental benefits started to blossom during the trial period. Rather, it was the concrete experiences – as individuals and in dialogue with others – that was new, about what

DOI: 10.1057/9781137276803

congestion fees means in practice in one's daily life. This was true for car users, among whom some reduced their car use in favour of public transport. It was also true for people who mainly used public transport' (Winslott-Hiselius, Brundell-Freij, Vagland, & Byström, 2009). The trial period was also a time for discussion among neighbors, colleagues, friends as well as following views in media. All of this was also important to the changes of attitudes, and in extension of norms, surrounding appropriate costs for car use in a congested area. This case illustrates how preferences are not fixed, and how habits, though demonstration and trial periods, in a small-scale and large-scale social contexts, may change into new, 'normal' practices.

If we take the potential of changes of social norms and values seriously, we should not be surprised if social norms concerning, for instance, tourist travel will change within the next few decades. This could entail norms through which distant trips to 'exotic' destinations are seen as old-fashioned and vulgar, after a period when the middle and upper middle class has been able to afford to travel 'everywhere'. Such norm change could become a ground either for higher taxation on such trips or for mandatory restrictions on individual air travel mileage. The main message here is that social norms, although having clear characteristics of being inert, are nevertheless volatile and changeable (in either a progressive or a regressive direction). Moreover, to the extent that various societal actors – governments, NGOs, companies – find it important, they may play a significant role in facilitating changes in norms.

Whereas changes in norms and practices may follow different sequences, a certain policy sequence appears to be particularly common. In his analysis of public awareness of 'the prisoners' dilemma' (such as acting towards the common good in a situation of uncertainty as to whether others do the same), Frey (2000) claims that carrots and (other) soft instruments usually are most efficient in the early phases of a norm change, by attracting citizen-consumers and companies to engage in the less environmentally harmful practices. Once citizen-consumers and companies have become accustomed to these new practices, and companies have developed technologies that further facilitate them, social norms may develop towards an acceptance of harder policy measures, such as fees and fines, for citizen-consumers and companies who stick, or move back, to the previous practices. The change from fossil-fuel-based to cleaner-fuelled automobiles or to public transport and bicycles can be

DOI: 10.1057/9781137276803

used as an example (Matthies, Klöckner, & Preissner, 2006). While earlier chapters of this book have pointed out the limitations of financial (dis) incentives, such instruments if used at the right place in a policy sequence have stimulated people to change their daily habits from daily car use to bicycling, which in turn has made them experience further advantages than the money savings. For example, people may experience health improvements, receive praise and appreciation from their peers, and may have experienced a positive response in that other people follow their example. Social motivation is strongly linked to all of these experiences.

Another example of a long-term policy sequence concerns norms and regulations of smoking. Chapter 3 discusses the Dionysian trust among certain groups of smokers. Still, one must admit that the dominant, social norm has indeed changed in certain parts of the world, at least to judge from the stricter regulations today than a few decades ago in the United States and the European Union. Crudely speaking, the dominant social norm there has changed along certain steps that at least the middle-age reader may easily recognize. A norm change such as this may not have occurred consciously, strategically, or through any long-term planning by anyone. These are roughly the steps in the sequence that I have identified (although not everyone would agree with the exact order):

- There are weak or non-existent prescriptions for how to smoke without disturbing others.
- It is polite not to blow out smoke directly on others.
- It is preferable not to smoke close to small children or close to people with breathing difficulties.
- Passive smoking is a health problem.
- Tobacco taxes are introduced and raised.
- Smokers should sit in designated areas in public spaces.
- Smoking is only permitted in closed, designated rooms or outdoors.
- Smoking indoors in public spaces, including restaurants and bars is forbidden.

In addition, there are signs of a normative change in that the smoking brake institutionalized in several countries has been removed at some workplaces, on the initiative of the management, unions, or through the influence of health organizations. Norms and policy instruments have become continuously stricter, and the public acceptance of stricter policies has become stronger. For instance, it should be noted that in social

DOI: 10.1057/9781137276803

interaction, it is often less awkward for a waitress to ask a smoking guest to put out his cigarette immediately, by referring to a law, than to ask the smoking guest to show a little more respect and consideration for her working conditions. It is not unlikely that norm changes related to other daily practices will pave the way for stricter policies. Acceptance of policy restrictions on certain types of cars, on driving or flight mileage, and meat consumption may very well change in the coming decades. Politicians with courage may introduce stricter environmental policies before the mainstream social norm is ready for it, which nevertheless may speed up this norm change.

Governments, companies, and interest organizations can facilitate normative change by learning from previous changes of social norms, such as the case with cigarettes. As to environmental harm, however, it is far from obvious that environmental arguments always constitute the most efficient problem framings for motivating citizen-consumers to help reduce environmental destruction. Concerning policy changes putting restrictions on daily car use, flights, and meat consumption, several other issue framings have been shown to be more motivating than climate change alone: health, local air, animal welfare, and as a basis the social motivation of strengthening the bonds between people in certain groups while marking distances to other groups.

5.2 Process is as important as outcome

In policy and planning with an explicit aim of reducing environmental harm, public participation is frequently brought in. Policy scholars and practitioners often mention public participation as crucial for a functioning and improving environmental policy and planning, albeit mainly at the regional and local level (Daley, 2007; Newig & Fritsch, 2009). This section is not intended to be a general subscription to local and public participation. There are hundreds of those already, some mainly ideological and others more analytical (for a theoretical and empirical overview, see Bäckstrand et al., 2010). Instead, the purpose here is to compare what would be the point and advantage or disadvantages of local public participation, if any, according to the three approaches analysed in this book.[1]

Subscribers to the ecological motivation approach would stress the importance of ecological and other knowledge exchange between the

DOI: 10.1057/9781137276803

local population (its diverse groups), entrepreneurs, ecologists, the municipality, and other actors. Local participation in environmental projects is worthwhile to the extent that it increases the knowledge of citizen-consumers, about the urgency of reducing environmental harms and about the ties between environmental harms – global, regional, and local – and local environmental efforts. Thus, local participation may stimulate citizen-consumers' potentially fundamental ecological values with acceptance and even support as a consequence. In windpower projects, the importance of open knowledge exchange is often stressed, for instance, as regards the consequences for birds, fish, noise, light, and aesthetics (Pasqualetti, 2011). The ecological motivation approach contends that if only the local population learns that the positive environmental role of a new planned windpower plant is stronger than its negative impact (if this can be shown as a fact), people can become motivated to support the project. In the spirit of this approach, researchers and entrepreneurs involved in windpower projects have tried to find general patterns and thresholds in terms of size, distance, noise level, and other 'substantive' factors, which distinguish local public acceptance from local opposition to the plants. Moreover, ways of presenting the environmental benefits of windpower in an understandable and concrete way to the local community are continuously being refined (Michaelowa, 2005). Conversely, entrepreneurs try to learn how to make use of the local ecological values in the communities, and to take these values into some account without having to change the initial plans of windpower plants too dramatically (Waldo, 2012).

Subscribers to the material motivation approach would also find local participation crucial for reaching solid agreement among stakeholders based on knowledge exchange and negotiations. Accordingly, local participation in 'environmental projects' has the potential of being useful to the extent that it strengthens citizen-consumers to get their due pay from local environmental efforts. Alternatively, it should be a right for the average person in the local community to influence the expenses of local projects and to have a say in how to minimize the costs. According to the material motivation approach, the main challenge of project entrepreneurs is to show that the local project is the optimal solution, in terms of economic costs and benefits from exploiting the local environment with windmills, compared with entirely different energy solutions and sites. Community participation may also put pressure on other stakeholders to give the local population as high an economic compensation as possible for any possible reduction in the attractiveness (such as tourism and economic land value) of the local area because of the exploitation. For citizen-consumers to have an impact on the outcome, on the final decision and its economic implications, is the *raison d'être* of public participation. In this approach, the local community's interests are likely to be fixed, oriented towards economic gain and the preservation of the scenery that people appreciate when moving to the area in the first place. Not in my backyard (NIMBY; Jones & Richard Eiser, 2010) combined with a view (sometimes shared with the previous ecological arguments) that windpower is not efficient enough as an energy source can often be heard in opposition to windpower plants. The Malmö region of Southern Sweden houses Lillgrund, one of Europe's largest windpower plants, and the entrepreneurs have worked hard to indicate the (societal), economic benefits of windpower to the local population:

> There are many myths about windpower, you know, that building windpower plants uses very much energy, so that there will never be any profit on it. And we try to punctuate those myths. Because after roughly half a year, the plant has produced the energy used in the construction phase. Windpower is competitive nowadays. (Waldo & Klintman, 2010, p. 17)

Even though the issue is not profitability for the local community, it is quite common that the general profitability of windpower is often debated back and forth between entrepreneurs and the community at stake.

On the other hand, endorsers of the social motivation approach may acknowledge important points in both of these other approaches. Yet, this third approach emphasizes the intrinsic value that people see in the decision-making process, sometimes regardless of the substantive outcome, the latter being factors such as where exactly the wind turbines are placed, how tall they are, and economic benefits to society. For the issue of local acceptance, the decision-making process even seem to be fairly independent of *the exact monetary level* of economic involvement the local community will have. People's substantive preferences (what is beautiful, natural, sound, and so on) are actually flexible to a larger extent than is often assumed, but the drive to strengthen or maintain one's social position through the 'dignity of decision making' is more essential to people. For instance, it is clear that that the level of acceptance differs substantially across countries and across local regions, differences that cannot be explained by the proportion of the landscape that has been used for wind power establishments or by other physical properties. Instead, the level of acceptance is rather connected to a range of factors

DOI: 10.1057/9781137276803

that concern the organizing and decision-making processes (Klintman & Waldo, 2008; Wolsink, 2010; Wüstenhagen, Wolsink & Bürer, 2007). The following quotation is probably typical among citizen-consumers who criticize a local windpower project:

> well, we are not against windpower plants as such, only against this establishment, since we felt run over [...] We felt that we were not allowed to say anything about it. And it's fair to say that we were faced with a *fait accompli.* (local citizen living close to the windpower plant Lillgrund, in Malmö; Waldo & Klintman, 2010)

What do scholars in evolutionary-oriented sciences say about public participation and deliberative decision-making? It is intriguing to note that they have for decades been very sceptical about the previously mentioned calls, calling them 'post-modern', 'social reductionist', 'ideological', and 'unscientific'. Recently, the fields of evolutionary psychology and cognitive science, in particular *the argumentative theory* school (Mercier & Sperber, 2011), have begun to claim the following: The primary, evolutionary function of reasoning, including scientific reasoning, is not substantive (about solving the 'physical' or 'concrete' problem at stake), but rather to improve one's social position. Moreover, this theory contends, groups from various backgrounds are much better than single individuals, including scientists, at solving the actual problems. This is because the group processes tend to 'even out' each individual's struggles to confirm his or her previous beliefs as well as each individual's attempt at convincing others for the sake of improving the persuader's social position (Johnson, 2011). Subscribers to the argumentative theory explicitly endorse the notion of deliberative democracy, maintaining that large groups interacting in an open debate climate are better than single individuals handling problems together (Mercier & Landemore, 2010). In defence of evolutionary psychology and cognitive sciences it should be stated that they are far more rigorous than the (other) social sciences in scrutinizing empirical evidence before making claims such as those in the argumentative theory. Together, the two former academic combatants' joint call for participation and deliberative processes for handling complex problems such as environmental ones is clearly a very powerful one.

However, based on the preceding findings in studies about argumentative theory, we should admit that the ecological and the material motivation approaches are right on at least one point: Public participation and

DOI: 10.1057/9781137276803

deliberative processes improve the quality of the knowledge that in sound processes may serve as a basis for the substantive outcome, such as of the physical and practical features of a windpower plant. If one subscribes to the social motivation approach from an evolutionary perspective, one has to admit that knowledge acquisition and sharing is an integral part of social motivation, albeit a sub-category.

In addition, the social motivation approach ought to admit that the material motivation approach is right in the claim that money matters. Perhaps the social motivation approach supported in this book has seemed to contend that money is more or less insignificant to citizen-consumers. Yet, this is far from the case. Particularly in societies marked by 'economism', which is the master frame in most societies worldwide today, material value is treated in public discourse as the basic human motivation. Many sectors and spheres of current societies use money as the major means to the deeper social motivation, including social status, bonding, and so forth. This treatment of the economic material value as a fundamental human motivation is frequently reflected in concrete policy and planning processes. For instance, in a comparative study of households in Småland (in Sweden) with households in Massachusetts (in the United States), economic arguments were given as the main rationale for all their choices, regardless of the heating system chosen: district heating (based on biomass), individual pellet stoves, oil, electric heating using heat pump, and geothermal energy. (Even a household of two 40-year-olds rationalized their continuing to use their old oil burner, by maintaining that they were 'too old' to invest in alternative heating equipment.) When people were questioned in more depth, however, it turned out that keys to their respective choices were issues such as trust in the national and municipal energy policy in the respective countries, the enjoyment of refining one's skills required by the different systems (mainly pellet stoves, but also heat pumps and the house equipment connected to district heating), and the exchange of experience between neighbours and friends. Lack of trust among the American respondents concerning government involvement in supporting renewable energy was a main theme. They claim that this lack of trust in energy policy has led them to try to become less dependent on the 'government – oil company cartel' that they perceived at the time of the study (in 2002).

> They [the government] are fools not to do more to promote substitutes and alternative renewable energy stuff: Wind, solar, biomass, you name it. It's

DOI: 10.1057/9781137276803

> really up to the alternative energy industry to make fools out of the old boy network. And show the population generally that you don't need to do this [use oil]. (Man, 50 years old, cook, lives in Somerville outside Boston with one adult. They have changed from oil burner to pellet stove as their main source of heating). (Klintman, Mårtensson, & Johansson, 2003, p. 39)

Interestingly, the view that the energy policy and planning of the government does not include enough renewable energy promotion has apparently been used as an argument among households for changing their heating systems to renewable energy. Moreover, in both Swedish and U.S. households using pellets for household heating, the joy of refining one's skills, and in some cases raising one's status to become some kind of 'independent', is mentioned as important, which may compensate for ambiguous economic calculations and considerably more practical effort.

> The pellets stove is actually much more work for me [than the oil boiler]. But I enjoy it, and it gets you that sense of freedom, whatever the hell that means' (Man, 50 years old, lawyer, living with one adult and two children outside Boston. They use a pellet stove, supplemented by an oil burner). (Klintman, Mårtensson & Johansson, 2003, p. 47)

Similar motivation can be found in municipal schemes of recycling and composting. For example, each household in a residential area in Ystad, Southern Sweden was as a consequence of a local policy lent composting equipment as part of the municipal policy to reduce municipal waste. After all the working- and middle-class households had described the practical efforts needed – to check the compost several times per day, make sure the temperature is right, add sawdust to get the right consistency, to deal with the risk of unwelcome guests (flies, wasps, rats) – they maintained that it was a good financial deal for them. Checking how much money they saved (by reducing their fees by producing less waste in the waste bin) the yearly savings amounted to a modest 50 euros per year in 2000 (Klintman, 2000).

Back to environmental policy and planning of windpower, the significance of financial 'carrots' is overwhelming. Scholars of local involvement and participation sometimes distinguish between three levels of participation: (1) informational, which mainly involves keeping the local community informed about the planning and development; (2) participatory, which refers to processes where the local public gets to have their say, and perhaps also to influence parts of the outcomes; (3) financial participation, which refers to cases where the local public may own shares

DOI: 10.1057/9781137276803

in the facility or be members of a windpower cooperative. Sørensen and colleagues, for instance, rank the third level, financial participation, as the most empowering mode of participation (Sorensen et al., 2002). This is indeed an important ranking, not least in development schemes with environmental ambitions in the global South, where 'community participation', orchestrated by organizations from the North often are promoted as progressive, although they only cover levels 1 and 2, failing to often involve small and medium enterprises (Klintman, 2012). Still, we have to distinguish between environmental projects in the global South, where parts of the local communities are faced with challenges of meeting their basic material needs, and those in the North that are developed under entirely different material circumstances. Consequently, with the social motivation approach, the actual amounts of financial involvement of the local population in the North ought to be perceived as of secondary importance for public acceptance, support, and participation in environmental planning projects. Although frequently discussed, the financial aspect is rather an integral part of a cluster of factors that make up the social motivation to participate. For local citizen-consumers to be invited to participate financially often seems to be highly motivating, not so much based on the actual sum that may be saved or earned, but because it raises the social status of local citizen-consumers, from spectators and discussion partners to actors whose social status in the project has become materialized.

5.3 The fundamental pillar of citizen activism

So far this chapter has to a large extent treated changes in environmental policy and planning as an issue of how policy makers, NGOs, and in some cases companies can stimulate citizen-consumers to *respond* to various policy instruments and 'do as they are told'. In policy making as well as in much research in environmental psychology, political science, planning, and marketing, this top-down perspective is the dominant one. It is certainly a very important one. However, as the sociologists Markham and van Koppen have convincingly shown in their edited volume on the history of nature protection in Europe and the United States, a top-down understanding alone would be not only over-simplistic, but plain wrong (Koppen & Markham, 2008). Sociology has long taught how bottom-up initiatives of citizen-consumers – from soft whistleblowing

DOI: 10.1057/9781137276803

to 'aggressive' activism – constitute the root of many current norms and regulations, concerning civil rights, women's rights, labour rights, and not least environmental problems. By examining the history of environmental policy making, it is clear that it was, and still is, efforts by citizens involved in grassroots groups – nowadays in collaboration with more or less professionalized NGOs– that pave the way for norm changes and environmental policy making. One of the founders of environmental sociology, Fred Buttel, has stated this order with the following claim:

> Environmental movements and activism are ultimately the most fundamental pillars of environmental reform. (Buttel, 2003, p. 306)

Note that Buttel uses the present tense. The role of environmental movements and activism remains crucial. It does so not only in the South, where scholars and NGOs often examine environmental grassroots activities, but also in the North, despite the so-called institutionalization of environmental issues in various regulatory frameworks. For example, this is reflected in the development of 'green consumerism'. Scholars and practitioners typically treat green consumerism as autonomous, conscious consumers – predominantly middle class, highly educated, white, urban women – choosing eco-labelled products and services in the store (for a critical discussion, see Boström & Klintman, 2009; Gilg, Barr, & Ford, 2005). However, green consumerism turns out to include a much wider set of tools through which people may act, ways which blur the border between citizenship and consumer roles. Together with Magnus Boström, I have presented this broader picture in Table 5.1.

TABLE 5.1 *Types of green political consumerism*

Green political consumerism	Monetary	Discursive
Frontstage	Consumers' boycotts or buycotts of products and services based on the green and political values of these consumers	Protests, demonstrations, and communication in media and Internet about products and services based on the green and political values of these consumers
Backstage	Consumers' monetary support to people and organizations that promote green political consumer issues	Consumers' participation, involvement, and debates in IGOs, NGOs, SMOs, and in public debates about political, consumer-oriented policy tools and practices

Source: Boström & Klintman, 2011, p. 39.

Supported by ample data, Table 5.1 indicates how monetary green consumerism at the frontstage (in stores) is highly dependent on support and engagement of citizen-consumers in the NGOs that develop the eco-labelling schemes. Moreover, these schemes, as well as environmental campaigns, eco-taxes, and hard regulation of harmful practices, would not have existed – and will not develop further – without the green political consumerism manifested in protests, boycotts, Internet mobilization, and so forth. In his study of environmental campaigns, Balsiger shows the importance of organizational efforts for green consumerism (Balsiger, 2010, pp. 325–326). His study actually demonstrates the interdependence, and in effect blurring of the distinction, between bottom-up and top-down as citizens engage in formal NGOs that in turn launch environmental campaigns.

5.3.1. Motivations for individual and collective action

Given the importance of bottom-up activities of citizen-consumers, particularly by supporting – or also participating in – collective action of NGOs, what are the driving forces of citizen-consumers, of the collective, and what makes governmental authorities or companies respond to these activities? From the ecological motivation approach, it is people's frustration with environmental deterioration and risks that is the basis for citizen engagement in environmental activism. A complex interaction between top-down and bottom-up takes place, as people are dependent on partly top-down initiated research and information about environmental risks and harm. Above a certain level of environmental harm and risks, people accordingly 'have had enough' and are motivated to engage in collective environmental action. According to this approach, to the extent that people assume that activism may trigger substantive reduction of environmental harm, this can be motivating for them. Social interaction and collaboration can thus be used as a means to the end of reduced environmental harm, which they believe may be clearly assessed. The ecological motivation approach assumes a more or less linear relationship between (perceived) environmental harm or risks and citizen-consumers' motivation for activism.

What would subscribers to the material motivation approach say about the driving force of bottom-up environmental activities? That approach shares parts of this linear view between perceived level of environmental harm and motivation to act. Yet, whereas the ecological motivation

DOI: 10.1057/9781137276803

approach conceives public awareness of global environmental risks and harm as a potentially intrinsic trigger of citizen-consumer mobilization, the material motivation approach implies an emphasis on people calculating trade-offs between collective engagement and perceived benefits; for example, in terms of reduced health risks to the individual or the person's near ones. Involvement in environmental issues requires investment of time, and this investment presupposes a conviction that the individual will benefit as directly as possible from this activity.

The social motivation approach shares certain features with these other approaches. The focus of the material motivation approach on direct benefits and risk reductions for the individual and her near ones is, in one respect, in line with what I have said so far about the evolutionary perspective to which this book subscribes: We exist today because those among the hunter-gatherers who had been provided through evolution with emotions giving priority to the handling of direct and short-term threats survived, and have been able to survive until this day. But this is only half the truth; those of the hunters-gatherers who survived did so because they were motivated to handle their threats *through social collaboration and distinction.* This social motivation is so powerful that 'intellectual understanding' of wider environmental threats should make us able to use our social motivation by joining collectives, and societal collaborative efforts, not just as a means to reduced threats, but as an intrinsic end.

It has become commonplace for social scientist to debate whether there is a decreasing or increasing interest among citizens in participating voluntarily in organizations and interest groups, whether the issue in question is to hit pins with a heavy ball or to enforce reductions of urban air pollution (Fischer, 2008; Putnam, 2001). This section does not aspire to analyse whether civil society activities, including environmental activism, are decreasing or increasing. Instead, I would like to argue that social motivation is independent of such possible changes in civil engagement. If civil society activities are decreasing (although I assume that they are decreasing in some parts and increasing in other parts), this should not be interpreted as a decrease in the components of the social motivation figure presented in this book's introduction (such as sharing, imitating, signalling, as well as managing problems in collaboration with others). Although even sociologists sometimes erroneously infer a reduction of social motivation in a culture from seemingly reduced activities in some parts of civil society, evolutionary theory convincingly shows that the former does not depend on the latter.

DOI: 10.1057/9781137276803

Thus, the passivity – omitted practical response – of citizen-consumers in policy issues should not automatically be seen as lack of societal and environmental concern. Some environmental sociologists and psychologists have tried to dig beneath the passive stance of citizen-consumers in certain, often local, environmental affairs. In addition to the impact of norms concerning whether activism and protests are 'appropriate', it turns out that a major reason for what may appear to be apathy is resignation, a sense of helplessness or futility, that single individuals cannot have any impact on policy processes (Wauters et al., 2010). This has been examined in several studies on climate adaptation (McLaughlin, 2011), policy processes concerning food safety in the European Union (Klintman & Kronsell, 2010), and criteria setting for eco-labelling schemes in the sectors of electricity and forestry (Klintman, 2009).

Such examples raise key issues of norms concerning democracy as well as of policy procedures discussed in Section 5.2.. However, it is both shallow and faulty to treat such 'apathy' as a trend towards reduced social motivation. Evolutionary theory demonstrates how and why social motivation is at the core of human nature: an integral part of human biology. Thus, the social motivation of citizen-consumers cannot be reduced; it can only be redirected to other channels of expression than those previously used. Our social motivation stimulates constant attempts by some individuals and groups to go against the current order, and to influence other groups and society as a whole to change its seemingly deterministic path.

5.3.2 Structure and (collective) agency

A principal claim of environmental sociology is that the citizen-consumer is not entirely free. Despite all the calls for us to change our individual lifestyles in a sustainable direction, the individual citizen-consumer is not autonomous. As Gould, Pellow, and Schnaiberg correctly observe, the individual choice of commuting by private automobile is highly dependent on city planning and other policies. Energy use in the household is dependent on what alternatives to oil are available on the market at reasonable prices, in turn possibly dependent on policy efforts (Gould, Pellow, & Schnaiberg, 2004).

In spite of such structural obstacles, there are very welcome conceptual tools for studying how structural change may take place. Transition theory provides such tools for examining opportunities and obstacles for transitions into more environmentally sound technologies and organizational

DOI: 10.1057/9781137276803

arrangements. Its main question is how less environmentally harmful 'niches' may develop into dominant 'regimes'. It is an undogmatic, analytical framework for understanding current or potential transitions into fossil-free modes of transport such as lowering chemical use in food production, more efficient recycling, energy systems, and so forth (Grin, Rotmans, & Schot, 2010). In its basic form, transition theory is fully compatible with evolutionary theory, through its acknowledgement of power struggles, inertia, and so forth, instead of providing naïve and rosy visions for new systems that can come about without human and political contestation.

Still, current structures do restrict our freedom as citizen-consumers. In our daily lives, we are highly dependent on – often large-scale – systems of provision for improvements of health and environment (Vliet, Spaargaren, & Oosterveer, 2010). The weight of structures and institutions, the prevailing 'order', is certainly where much of these environmental challenges (and possible solutions) need to be investigated. Evolutionary theory would echo this. From an evolutionary perspective, it is fair to assume that governments, and to some extent companies have an interest in maintaining important parts of status quo, with fairly small adjustments. Governments and the big corporations (including the leaders of these 'organisms') are by definition at the top of the social hierarchy. Comprehensive change is likely to be perceived by them as jeopardizing their high social position. The social motivation of agents in organizations at the top points in several ways towards efforts at maintaining the prevailing order – again, with changes at the margin. As we have seen in earlier chapters, social bonding and 'energy efficiency' through norms, habits, routines, and rituals is part of human nature; this human trait can be extrapolated to an interest among leading organizations and agents in preserving current norms, procedures, and power relations. In particular, those at the top would have little social motivation for making their dominant position more uncertain and unstable (which is what processes towards comprehensive policy change necessarily entail).

Several sociologists share this understanding of societal change with evolutionary scholars, although the two groups rarely refer to each other. For instance, when sketching what they call their 'general theory of strategic action fields', the sociologists Neil Fligstein and Doug McAdam describe this issue of status quo:

> Incumbent socially skilled actors will defend the status quo. It follows that if a new frame emerges, it will come from an invader or challenger groups. They will attempt to create new rules and a new order and therefore either will

DOI: 10.1057/9781137276803

> build a new political coalition based on interest or create a new cultural frame that reorganizes interests and identities. (Fligstein & McAdam, 2011, p. 18)

The role of citizen-consumers in environmental issues is a good illustration of the leading societal segment's ideal at the top of the status quo with minor adjustments. Whereas imperatives are often heard from state agencies and companies about citizen-consumers' obligation or opportunity to reduce environmental harm, these leading bodies exercise cross-pressure on citizen-consumers, for instance, by simultaneously problematizing reductions in Christmas sales, automobile sales, and so forth. To go far beyond light green consumerism (choosing eco-labelled products, recycling, and using cars a bit less, but continuing 'normal' patterns of consumption) seems in general to be beyond the comfort zone of governments.

The general comfort of the status quo among the leading segment of society is interpreted by the material motivation approach as a fixed preference among the leading segment for the *substantive content* of the current order. But, for example, is a high number of cars on the roads or extensive transport of goods with trucks rather than with trains a fixed preference of agents in this high segment? Leaving aside the enormous issue aside of lobbying by industries (such as car manufacturers or oil companies), the governmental interest in preserving the status quo is not the same as perceiving the current order as optimal per se. As shown earlier in this book, when distinguishing between *forms* of values (e.g., ethical and metaphysical values), which are part of human universals, and value *content* (e.g., whether a person's flights per year above a certain numbers are considered unethical), differs vastly across groups, cultures, and times. Similarly, preference of the status quo is *form*, a general preference for stability and certainty among the leading segment that they will remain at the top. Still, this only means that in order to achieve comprehensive policy change, for instance stimulating reduced environmental harm, we should not expect comprehensive initiatives from the leading segment of society. Instead, efforts are needed from 'below', from citizen-consumers and their activities in civil society. In the long run, the leading segment may not be against such changes once they have become reality like civil rights. Yet, this requires much skill from civil society actors. The previously mentioned sociologists Fligstein & McAdam are again in line with evolutionary theory:

> Proposition 2. Skilled social actors are pivotal for new fields [units of collective action] to emerge. They must find a way to translate existing

> rules and resources into the production of local orders by convincing their supporters to cooperate and finding means of accommodation with other groups.
>
> Proposition 3. Skilled social actors can help produce entirely new cultural frames for fields. They do so by building compromise identities that bring many groups along. In this process, every group's identities and interests can be transformed. (Fligstein & McAdam, 2011, p. 6)

Despite human inertia, habit-oriented social norms, and the ruling segment's strong interest in maintaining the status quo, there will always be a social motivation for individuals below the dominant segment to try to challenge status quo. This dynamic between the prevailing order and attempts by some individuals to develop an alternative to improve their own position is very much in line with evolutionary theory in several of its disciplines. Environmental issues, not least *vis-à-vis* the roles of citizen-consumers, are likely to be subject to continuous attempts at creating alternatives to the status quo, and to attempts at integrating reduced environmental harm into the current market-economy system (with many different varieties of eco-taxes and degrees of 'progressive' versions). Still, we should bear in mind the sociological assertion that separate individuals 'at the bottom' can do little to change consumer society's massive use of resources (unless such changes have already been incorporated into policy schemes). The issue of group formation becomes crucial here, as already described by Fligstein & McAdam. Group formation is the core of our social motivation, both intrinsically (we are emotionally driven to form and join groups because it makes us feel good) and instrumentally (group participation is in general the best way to solve problems). Whereas individuals are often hesitant about protesting against, for instance polluting companies, the inclination to protest increases significantly if individuals are or become part of collective protest efforts. A study from China, Liu and colleagues indicate how collective formations increase the likelihood of protest against polluting companies (Liu et al., 2010, p. 1001). In addition to the instrumental advantage (of increased power) that citizens may perceive when joining a collective, the more intrinsic social benefits of uniting in collective protest action have been corroborated in several studies. The variable 'subjective norm', often scoring high in motivational studies of environmental activities (including political activism) of citizen-consumers, are founded on '"individuals" perception of whether other important people in their life would want them to perform the behavior'(Fielding, McDonald, & Louis, 2008, p. 319).

DOI: 10.1057/9781137276803

As I have already mentioned, environmental groups operate under various degrees of radicalism. Certain NGOs developing eco-labels for goods and services create these schemes through consensus-oriented collaboration with industry and governments based on an eco-pragmatic framing of the problems at stake (Klintman & Boström, 2004). Other groups tend more to be activists. For instance, scholars have observed a fairly common frontline position of women, not least in the role of mothers, in environmental activism. For instance, Nobel Prize winner Wangari Maathi initiated Kenya's Green Belt Movement; moreover, mothers in the Chipko group created a movement testifying in global forums about environmental injustices leading to severe birth defects in the countries they represent (Logsdon-Conradsen & Allred, 2010). At the end of the scale of radical activism, there are the militant environmental groups, such as Earth First, the Animal Liberation Front, and Earth Liberation Front. It is common to see environmental groups of various degree of radicalism as competitors only. Both social movement studies and evolutionary research emphasize the importance of group distinction. However, recent studies have shown how radical environmental groups are often very far away from the impulsive eco-lunatics they are typically portrayed as in mainstream media (Delingpole, 2012). Instead, regardless of whether one approves of their willingness to damage property or not, one must admit that they are more sophisticated in terms of strategy than is often generally understood. It turns out that these groups have sometimes worked with the main aim of having 'mainstream, light green' environmentalism and policy processes move in a greener direction, hence making the dominant social norm of 'normality' greener:

> Insiders have acknowledged that the formation of the well-known direct action group Earth First! was an intentional strategy by activists 'to allow the mainstream groups to look less radical and achieve more protection for the environment'. (Goodman & Sanders, 2011, p. 139; quoted in Scarce, 2005, p. 6)

This sheds light on the complex dynamic between environmental groups, and how they may actually be opposed to each other's ideology, while still reinforcing one another.

5.3.3 Collective action beyond the local level

A final remark about collective environmental action by citizen-consumers concerns geography and size. To be sure, evolutionary theory is often based on empirical research and theorizing about local social settings similar

DOI: 10.1057/9781137276803

to our ancestral environment of hunter-gatherer society. Nevertheless, it is important to note that the social motivation approach influenced by evolutionary theory is far from confined to social bonds and distinction across groups of the size of those in our ancestral environment. Instead, humans turn out to have an amazing talent for acknowledging practices, tastes, and skills of remote people of the same, or slightly higher, socio-economic status. Ever since the dawn of transportation and migration (in prehistoric times), people have been inclined to develop 'weak ties' to people at a certain distance. Evolutionary theory uses the term 'reciprocal altruism' to explain the basic social motivation for collective action that goes far beyond people of our own kin with whom we share many of our genes (Trivers, 1971).

Albeit separate from evolutionary theory, the sociologist Mark Granovetter is well known for emphasizing – in theory and repeatedly confirmed in empirical network studies – the strength of social ties in terms of knowledge acquisition, the potential for success on the labour market, and so forth (Granovetter, 2005). Others have indicated the importance of weak ties for the success of social movements, such as in handling environmental issues. By reflecting on the parameters of the social motivation approach presented earlier in this book – such as the development of skills and knowledge to share and signal as well as to imitate and diverge – in the light of cultural development, it is clear that much of the development has taken place through weak ties, not least in intercultural regions. This explains the internationalization of environmental organizations, both the more activist kind (Van Der Heijden, 2006) and reform-oriented, eco-pragmatic organizations working towards international harmonization of eco-standards (Tamm Hallström & Boström, 2010). Overall, citizen-consumers are crucial for both kinds of collective action, as participants, supporters, and users of the policies and schemes these groups help develop.

5.4 Conclusions and discussion

This chapter indicated the importance of understanding citizen-consumers not just as choosers of goods and services in the store, but also as responders or initiators of policy and planning measures with vast environmental implications – both positive and negative ones. The section examining citizen-consumers' readiness to accept environmentally

DOI: 10.1057/9781137276803

promoted policy schemes criticised attitude surveys that are based on the material motivation approach, surveys implying that people's preferences are fixed and oriented towards comfort as well as economic gain. Moreover, the ecological motivation approach was criticized for placing too much confidence in environmental facts as a potential catalyst of preferences for public acceptance of environmentally oriented policy and planning. By presenting empirical cases of how initial public scepticism has given way to public acceptance, for instance in favour of congestion fees in Stockholm, this section highlighted how acceptance can grow out of factors that are part of social motivation: pilots, trial periods, and so forth, which integrate the policies with people's development of practical skills and habits, as a rule in dialogue with their peers until the new routines have become part of social norms.

The second section elaborated on the frequently mentioned importance of public participation in environmental policy and planning. What would be the advantages or disadvantages of public participation, if any, according to the three approaches analysed in this book? Based on primary as well as secondary sources of research on local windpower opposition and approval, this section exemplified and illustrated the various arguments. The ecological motivation approach would stress the importance of learning about specific ecological conditions in the local areas, and of spreading knowledge about the benefits of wind turbines (weighed against environmental risks to bird life, aesthetics, and so on). The material motivation approach would find local participation crucial for reaching solid agreement among stakeholders based on negotiations. Accordingly, the local community's interests are likely to be fixed and oriented towards economic gain, and the conservation of the scenery that led people to move to the area in the first place. The social motivation approach sees important points in both these other approaches. Yet, based on recent developments in argumentative theory (with an evolutionary foundation) as well as on sociological studies of participation, this third approach emphasizes the intrinsic value that people hold in the decision-making process itself, sometimes overriding the substantive outcome, such as where exactly the wind turbines are placed, how tall they are, how much economic involvement the local community will have, and so forth. The social roles that the local citizen-consumers are given, or are able to take, are intimately related to their social motivation, and are thus as a rule more important than the physical outcomes. As to the monetary interests of the local population in cases of windpower, it

DOI: 10.1057/9781137276803

turns out that the role as co-owners that some schemes invite the local population to adopt is crucial, whereas the amounts are secondary. In the third section, this chapter indicated how the social motivation is involved in schemes initiated with citizen-consumers as part of civil society. In fact, this is the very root of today's environmental politics and policymaking. The section showed the enormous force that social motivation constitutes for more progressive environmental efforts in society: through protests, boycotts, and so forth. However, a point not covered earlier in this chapter is the following: When acknowledging the fundamental role of social motivation for participation and activism, one must also take into full account the potential dynamics of group attraction and distance; sub-cultures can easily evolve that mark their social position by making counter-efforts, either in the name of the environment (as an anti-windpower lobby) or as something else (apocalyptic religious groups). Still, as indicated by evolutionary theory and research on action fields or social movement studies, a more fundamental issue is how the leading political and industrial segment of society could become motivated to risk their social position by inviting environmental activism to seriously challenge the current, environmentally harmful state of affairs. Moreover, as the next chapter illustrates, while wider societal change is possible, it cannot be designed in just any way that social and environmental thinkers may wish.

Note

1 It should be mentioned that all three approaches are consistent not just with praise, but also with certain criticism of public participation. Moreover, all approaches may embed the belief that direct public participation should be limited, for instance where local participation is undemocratic, locally biased, and involving local elites who make claims in the name of the 'community', and so forth. Also, as was seen Chapter 3, the ecological as well as the material motivation approaches include a strong belief in authority, experts, and scientific facts, factors that are partly at odds with a subscription to participatory democracy.

DOI: 10.1057/9781137276803

6
Wider Societal Change

Abstract: *The chapter analyses three recent green visions of social thinkers: ecological citizenship, alternative hedonism, and degrowth. It tries to evaluate these visions based on the extent to which they are coherent with various aspects of social motivation. In this respect, all three visions show strengths and weaknesses.*

Klintman, Mikael. *Citizen-Consumers and Evolution: Reducing Environmental Harm through Our Social Motivation*. Basingstoke: Palgrave Macmillan, 2013.
DOI: 10.1057/9781137276803.

After the preceding chapter's focus on citizen-consumers' readiness to be proactive or reactive about single policy changes in environmentally sound directions, this chapter broadens the scope. What can be said about citizen-consumers' current and future willingness to initiate or support changes that move beyond single policies towards wider societal change? Such changes may need to include what Eckersley calls changes in institutional arrangements and policy paradigms; that is, changes within wider social norms, goals, goal hierarchies, political priorities, and so forth (Eckersley, 2004). Some researchers, including Eckersley, maintain that any state wishing to call itself 'green', by becoming 'climate-neutral', needs to alter its policy paradigm.

Previous studies have considered this issue mainly in terms of ideology or of views on nature among people and organizations: 'To what extent are people willing to lower their standard of living?' is a question asked by those assuming that environmental adaptation must entail a regression to less (materially) comfortable ways of life. Others, such as the ecological modernization programme and theory, portray and support a smoother transition to a 'deepened' modernity, more technologically advanced and comfortable, while at the same time 'smarter' in its reduced resource use. The role of citizen-consumers, from the latter perspective, is to make use of their liberal empowerment, by choosing 'green' products and services, and supporting moderate policy reform (Mol & Spaargaren, 2004). A third school conceives the entire environmental rhetoric as a construction by groups domineering society, with these groups, like priests and ministers, prescribing substantial changes in the way of life of the general public, while simultaneously maintaining and increasing their own use of resource-depleting products and services (Kilian & Elgström, 2010).

This chapter discusses and comments on three different – but partly overlapping – suggestions for an alternative society: ecological citizenship, degrowth, and alternative hedonism. I analyse the suggestions from the three perspectives that have been used throughout the book, ending with a more extensive discussion of what a social motivation approach can contribute to the three visions, with particular focus on citizen-consumers.

6.1 Alternative hedonism

Alternative hedonism is an empirical acknowledgement as well as a normative suggestion advanced by the philosopher Kate Soper. In contrast to many moralist propositions under the ecological motivation

DOI: 10.1057/9781137276803

approach, Soper emphasizes that consumerist lifestyles are to a large extent illusorily hedonistic (Jackson, 2005). Although distancing her perspective strongly from the criticism of being 'anti-modern', she reverses the common view of hedonism by claiming that modern affluent society in many respects seems 'overly ascetic and sensorily deprived' compared with earlier times. Some of the downsides of mass consumerism, not just for the environment but for self-interested individuals, are the following, according to Soper:

> Consumerism is today for many people both compromised by the pollution, congestion, stress, noise, ill health, loss of community and personal forms of contact it entails, and viewed as pre-emptive of a distinctive range of pleasures. (Soper, 2008, p. 571)

In Soper's view, the downsides to mass consumption are so numerous and extensive – even at the individual level and even for affluent consumers – that it would be more pleasurable, even to self-interested, middle-class Euro-American consumers, to move their lifestyles in another direction: towards alternative hedonism. In Levett's formulation, this would lead to 'the sensual pleasures of consuming differently' (Levett 2003: 60–61). An anti-consumerist ethic and politics should therefore appeal not only to altruistic compassion and environmental concern but also to the more self-regarding gratifications of consuming differently: to a new erotics of consumption or hedonist imaginary (Soper, 2008, p. 571).

What would be examples of alternative hedonism? Soper's main example, or basis, is 'freedom from the car-culture depredations', a freedom allowing for multiple cycle routes, green cities, less step-motherly treatment of pedestrians, and expansion of the railway. Shortening of the working week would be part of alternative hedonism, as would better insulated homes, luxurious public art, more cultural education for its own sake, rather than as instruments for economic growth. Although this sounds very much like the wish list of any environmentally oriented academic citizen-consumer of mature age (Soper and myself included), the list could be expanded to include urban planning and better, less environmentally harmful access to outdoor life.

A major virtue of the notion, and empirical identification, of alternative hedonism is that it does not refer to a project of trying to redesign our human nature into one of (unreciprocal) altruists. On the contrary, the main argument is one of individual hedonism, by emphasizing the

DOI: 10.1057/9781137276803

importance of 'self-regarding gratifications'. In this respect, it overlaps the prescription in Chapter 4 that policy makers should not treat environmental problems as *environmental* only, but also as health problems, economic problems, problems of excessive time spent commuting on congested highways, problems of working conditions, crime, and so forth. Yet, the alternative hedonism suggestion ignores parts of the social dimension of human nature, particularly competition. Alternative hedonism fails to acknowledge the strength of a major motivation behind consumerism: the motivation for social status. It is this motivation that has us work more than we would need if we were entirely 'materially rational', only craving the products and services required to support our basic material needs. Thus, one should also acknowledge the actual happiness that emerges in us when we have purchased a beautiful, well-functioning item, through which we can signal our social position to ourselves and to others. It is difficult for a person to hide his joyful Dionysian trust in his new cabriolet on an early summer ride. Perhaps it is mainly when consumerism becomes an *inefficient* instrument of social signalling that the citizen-consumer occasionally – as most of us have – gets the feeling of 'emptiness' after having unpacked our ten shopping bags at home. The idea of alternative hedonism would need to show how it can involve the social game in a less destructive way than conventional mass consumerism.

According to the alternative hedonist approach, creative work and leisure entails sufficient potential benefits for the social dimensions of our lives, such as more time with family, friends, and so forth. Yet it ignores the social game, and the essential parts of the social game of which consumerist lifestyles are instruments in today's society. Thus, the post-modern theories, counter to what Soper asserts, are right when they point towards the essentiality of self-expression and so on in consumer society. From a Veblenian as well as an evolutionary perspective, we all do quite hard aesthetic work in our consumerist lifestyles, as a means towards doing well socially. It would be a shallow understanding that we just buy endless amounts of items with the value reflected in the slogan placed on the T-shirt: 'He who dies with the most toys wins.' Instead, what may be seen as excessive amounts of stuff from the perspective of natural resources and global welfare takes the shape in daily life of aesthetic selections, efforts to be creative, and to develop our own style in order 'to do well in the social game' through cultural distinction and attraction. As everyone knows, this is a delicate balancing act, because we typically purchase quite

DOI: 10.1057/9781137276803

similar products and services in the same stores during the same periods. When thinking about less harmful alternatives to mass consumerism, it is highly important to take the basic 'evolutionary' dimension of mass consumerism seriously, namely the social motivation, which includes the factors that are shown in Figure 1.1 to imitate and diverge, share and signal, as well as develop skills and routines.

In order to find reasonable alternatives for wider societal change, these alternatives must satisfy the same evolutionary motivations, avoiding practices that members of society agree are destructive. Thus, consumer culture is perhaps not a hedonist culture that can just be changed into an alternative hedonism, but instead one way for citizen-consumers struggling to meet their social motivations – through hard, aesthetic, ethical, economic reasoning, and work – that might need to be replaced by another playing field for these struggles, one that is less destructive to the environment and to global welfare. The reason for downplaying these social struggles in the alternative hedonism position might be the partial Marxist foundation of the alternative hedonism that makes it more similar to the utopia of classless society than to the envisioning of a society that takes the overriding social values into consideration. From the latter position, taking the social motivation seriously would entail policy-related and cultural activities, in collaboration with citizen-consumers, aimed at moving necessary competitive traits to another battlefield than conventional consumerism. This playing field may very well be post- or anti-consumerism.

Instead, alternative hedonism has so far been formulated as a prescription for how to turn us into advanced post-consumerist noble savages, interested in developing our crafts and skills as an intrinsic drive, with little or no urge to compete with and to defeat others, in skill, generosity, and so forth. An overwhelming number of studies from indigenous, rural, and urban societies show social status as an underlying drive along with collaboration, and the noble uncompetitive savage, contrary to the claims of the influential social anthropologists Margaret Mead and France Boaz (Mead, 1928), has probably never existed (Freeman, 1999). If she had, she would have become extinct early in our ancestral environment (Wilson, 2004).

To be sure, alternative *hedonism* seems to harmonize quite well with the Dionysian dimension that this book claims to be as important as the extent to which it has been forgotten. What is appealing about the alternative hedonist concept is Soper's ambition to take feelings and

DOI: 10.1057/9781137276803

pleasures seriously. Still, she only takes half of our feelings and pleasures seriously, namely the more Apollonian dimension of our motivation (which we admit is part of human nature): craft, construction, refinement, and so forth. Whereas an alternative hedonism is presented by the use of the term 'new erotics of consumption' (Soper, 2008, p. 571), alternative hedonism does not yet include what humans typically see as the most powerful erotics: dynamics in human relationships. Instead, it has so far been depicted as mainly auto-erotic or social in an un-Dionysian manner.

6.2 Ecological citizenship

Advanced by Andrew Dobson, the notion of ecological citizenship is a prescription: 'The principal ecological citizenship obligation is to ensure that ecological footprints make a sustainable, rather than an unsustainable, impact' (Dobson, 2004, p. 12). The ecological footprint is in turn an image and an educational tool through which we may compare our own negative environmental impact (such as emissions through our transportation and choices of goods) with other people, either in our own society or in other societies (Wackernagel & Rees, 1996).

As opposed to alternative hedonism, with its empirical ambition of indicating what *true* hedonism would be, ecological citizenship is instead a contribution to normative political philosophy. The notion of ecological citizenship seems to provide politicians, policy makers, and NGOs with conceptual tools and arguments for constructing policies to reduce environmental harm (at the macro level). At the same time ecological citizenship has traits of *moral* philosophy addressing citizen-consumers more directly, telling us directly what we ought to do, why we should do it, and how we should assess the level of success.

Ecological citizenship is best understood by elucidating its ties to classical political theory. A classical division is the one between liberal and republican citizenship (N.B. with little in common with contemporary US politics). The former has its focus on citizens' rights – to social security, to vote, and so forth. According to Dobson, this preoccupation with rights needs to be combined with, or partly replaced by, the other part of this division, namely republican citizenship (Dobson, 2007, p. 280). This latter type of citizenship focuses instead on citizens' duties and virtues (Trentmann, 2007, p. 148). To be sure, political normative theorists often

DOI: 10.1057/9781137276803

find it beyond their scope to consider the issue of how people can become motivated to follow the theorists' imperatives. Dobson is an exception. According to him, people's knowledge about their environmental harm, particularly the asymmetrical harm caused by people in the North, should motivate them to reduce their ecological footprints.

> The citizen that sorts her garbage or that prefers ecological goods will often do this because she feels committed to ecological values and ends. The citizen may not, that is, act in sustainable ways solely out of economic or practical incentives: people sometimes choose to do good for other reasons than fear (of punishment or loss) or desire (for economic rewards or social status). (Dobson, 2007, p. 279)

Agreeing with Beckman (2001, p. 179), Dobson maintains that 'people sometimes do good because they want to be virtuous' (Dobson, 2007, p. 279). In the terms of this book, the notion of ecological citizenship strongly criticizes the material motivation approach, by emphasizing that human motivation need not concern economic reward or practical advantages. Instead, an ecological motivation appears to be sufficient for citizen-consumers to make extensive changes in their lives. Interpreting Dobson in terms of the 'is'-'ought' separation introduced in 2.2.1, both an 'is' (environmental deterioration), is obvious to Dobson, and an 'ought', the practices among citizen-consumers that would reduce this deterioration. This state of affairs may, according to his notion of ecological citizenship, lead citizen-consumers directly to a deep motivation to change their practices in this virtuous manner. Dobson states his prescription while claiming that he tries 'to avoid a *naïve* voluntarism' (Dobson, 2007, p. 277). A couple of examples of what these practices are have been mentioned already, such as purchasing ecological goods and sorting garbage. Several other scholars support the notion of ecological citizenship by extending the list of possible empirical areas where it could be practised (about climate change and personal carbon allowances, see Jagers, 2009; Wolf, Brown, & Conway, 2009). For instance, Jill Seyfang, who closely examined a local organic food network and its members, declares that she saw several signs of ecological citizenship as a driving force among the members. She holds that

> the values and principles expressed by both creators and users of this local organic food network are strongly resonant with ecological citizenship, and a strong environmental ethic is a significant – if not primary – motivation for many of the participants. (Seyfang, 2006, p. 393)

DOI: 10.1057/9781137276803

Also, according to Seyfang, ecological citizenship as a basic driving force was to a large extent manifested in the way people discussed their motivations and concerns:

> Indeed, many participants used the language and vocabulary of ecological citizenship when explaining their motivations: reducing ecological footprints and cutting consumption were commonly cited, in addition to more personal health and safety reasons. (Seyfang, 2006, p. 393)

In this and several other empirical studies, the normative component of what good, virtuous citizens ought to do is mixed with empirical claims of 'opportunities' of citizen-consumers. But what is really shown in this and other empirical studies? Is it really that environmental ethics is the basis here, as the prescribers of ecological citizenship seem to maintain? In Chapter 2, I distinguished between values and forms of value, arguing that the latter are universal and thus basic: aesthetic, ethical, metaphysical, and material. For instance, The ethical value form is universal: People in all cultures and societies hold ethical values. Yet, the value forms leave much space for each culture to specify the content of this value form, what would be considered 'good', 'virtuous', and so forth. The most basic motivation for acting upon this goodness or virtue, or for opposing it, needs to be social.

Intriguingly, and consistent with the evolutionary based social motivation approach advanced in this book, Seyfang briefly admits that the organic network in question (called Eostre) 'uses food as a mechanism for community-building and social cohesion' (Seyfang, 2006, p. 393). Based on the social motivation approach, it is clear that the work of this community is fully dependent on the social motivation of its members. What has happened is that the social norm of ecological citizenship has been made consonant with the social motivation of these people.

We should also note that an ecological motivation (were it independent of social motivation) would plausibly entail ecologically based questioning of the current practices of the community: Are localized food systems always consistent with minimization of ecological footprints? Is organic food production unquestionably the only possible way forward for meeting the larger ecological and health-related challenges facing agriculture? Instead, the community settles with this framing of ecological soundness, a consensus that is effective for maintaining and strengthening social identity. Such a framing helps to simplify a complex world and strengthen social bonds, yet perhaps at the expense of an 'ecological optimization'.

DOI: 10.1057/9781137276803

The normative theory of ecological citizenship may certainly be useful for studying groups and communities that have used this framework in their expressed values and rationales. Moreover, ecological citizenship may provide a basis for politicians', policy makers', and companies' framings, as they try to develop schemes for motivating various groups of the public to reduce environmental harm. However, to become deeply motivating for a large part of the public, the social factors addressed by evolutionary theory must be integrated into the framework. How can our need for social signalling and knowledge sharing, information, and skills be incorporated into schemes aimed at ecological citizenship? How can we satisfy our need, not only to imitate other people's 'green' practices, but also to diverge and distinguish ourselves from 'those other' ecologically oriented groups? How can the need to 'show off' be used concerning some community members' revised knowledge and refined skills about how to bring about constant ecological improvement of the routines of the community?

To become fully useful, the notion of ecological citizenship needs to fill the sociological and evolution-based gap that such questions represent, between political normative theory and micro-oriented, case-based evidence of the *stated* ecological rationales among members of single communities. Filling this gap is not at all impossible, but it requires conscious efforts among scholars and practitioners to make ecological citizenship compatible with sociology and evolutionary theory. To be fair, it should be noted that certain signs of this can be found in Dobson's own reasoning when, in certain parts of his work, he abandons his belief in an intrinsic ecological motivation. For instance, he quotes Jackson, correctly claiming,

> that there are some strong suggestions that participatory community-based processes could offer effective avenues for exploring pro-environmental and prosocial behavioural change. (Jackson, 2005, p. 133)

6.3 Degrowth

Degrowth has been defined as

> an equitable downscaling of production and consumption that increases human well-being and enhances ecological conditions at the local and global level, in the short and long term. (Schneider, Kallis, & Martinez-Alier, 2010, p. 511)

In order not to worry people living within the paradigm, or master frame, of economic growth, degrowth proponents distinguish between

DOI: 10.1057/9781137276803

'unplanned degrowth within a growth regime', which would be depression, and 'a voluntary, smooth and equitable transition to a regime of lower production and consumption', which would be sustainable degrowth (Schneider, Kallis, & Martinez-Alier, 2010, p. 512). According to the degrowth scholars Odum and Odum, such a transition would constitute a collective management of a 'prosperous way down' (2001).

The influences of the degrowth notion are wide and the group of proponents is eclectic. The choice of the term 'economism', and not 'capitalism,' to explain the societal paradigm that degrowth proponents criticize, reflects the efforts not to be placed on the left wing on a conventional political scale. Historically, influences have come from, among many others, Schumacher's famous book, *Small is Beautiful* (1973), denouncing growth, economic accumulation, large-scale industry, and centralization. A main field influencing degrowth theory is ecological economics. One of the persons held up as a founding father is Georgescu-Roegen, initiating the notion with his paper 'Energy and Economic Myths' (Georgescu-Roegen, 1975).

What are the concrete changes that subscribers to degrowth would like to see? Among several other changes, I only mention the ones most relevant to this book:

- Dematerialization, the transaction of more services and fewer goods, although degrowth proponents are far from alone in this respect (Lorek & Fuchs, 2011).
- Limits to what is regulated with money. For instance, there is scepticism concerning increased energy and resource prices, because it is claimed that such increased prices run the risk of harming the less affluent parts of the population. Still, there is a pragmatic tendency in the degrowth proponents, placing some hope to, for instance, CO_2 taxes (see Kallis, 2011, p. 874).
- To strengthen local economies, through decentralization of financial institutions and banks, through local complementary currencies, and so forth (Korten, 2009).
- Shorter working hours have been proposed and discussed extensively, on the basis of calculations of GDP, 'genuine progress indicator' (GPI), greenhouse gas emissions, and so forth (Victor, 2010).

Since degrowth proponents criticize the core of modern industrial society, namely economic growth, directly, degrowth has been subject

DOI: 10.1057/9781137276803

to more criticism and attacks than have alternative hedonism and ecological citizenship. Aside from the predictable criticism from growth proponents (that degrowth is either unrealistic, or that it would of necessity generate economic depression), criticism of degrowth by scholars more sympathetic to the idea refers to the vagueness about what it would actually mean to make society degrow, and what concrete policies would work in this direction. Moreover, critics contend that there is not full consensus as to whether degrowth would be consistent with capitalism. In addition, it is not always clear how the notion of degrowth is different from the idea of sustainable consumption (Van Den Bergh, 2011). It is beyond my scope to shed light on the whole variety of criticism and defence of degrowth. Rather, the task here is to analyse to what extent degrowth would work given lessons of social science and evolutionary theory, focusing on human universals and social motivation. To what extent would degrowth be compatible with lessons from evolutionary theory, particularly the social motivation approach, with the focus on environmental impact?

First of all, it deserves to be mentioned that economic growth has been problematized from many directions, not only from 'dark greens' and leftists. For instance, the neo-classical economist and Nobel laureate Robert Solow, with a particular interest in growth, has admitted that

> either continued growth will be too destructive to the environment and they [the United States and Europe] are too dependent on scarce natural resources, or that they would rather use increasing productivity in the form of leisure. (in Stock, 2008)

Furthermore, for many thinkers with an evolutionary outlook, the idea of doing something about economic growth, because of its possible downsides, is far from outrageous. Robert Wright, an evolutionary thinker and writer with a market-liberal perspective, acknowledges the problem of 'cultural lag', namely that cultural and social institutions (social norms, regulation, and so on) lag behind the pace of technological and chemical advancement, in turn closely tied to economic growth. Whereas the sociologist Ogburn, who coined the term 'cultural lag' called for speeding up social institutions (Ogburn, 1922), Wright, well aware of the evolutionary based relative slowness of changes of norms and other social institutions, instead prescribes that the *pace* of technology development, and thus economic growth, should be controlled and slowed down 'just a notch' (Wright, 2001, pp. 232–233).

DOI: 10.1057/9781137276803

A highly relevant risk with economic growth that degrowth theorists address is the risk of a rebound effect, which was discussed in Chapter 4. If economic carrots and sticks are the main instruments used to have citizen-consumers economize on energy and other resources, there is a high risk that the money saved will be consumed on increased consumption, such as longer commuting distances, a higher number of cars, and more tourist air travel. Degrowth theorists make important contributions to the wide environmental debate by emphasizing the need to go beyond efficiency improvement when developing schemes for reductions of citizen-consumers' environmental harm.

What if citizen-consumers and their motivations engage in the 'voluntary simplicity', that degrowth theorists call for? As the reader probably has observed, this book and degrowth theory share the view that material accumulation, beyond the material standard to meet basic human needs, does not necessarily enrich our lives in any direct proportion to the increases in material standard. Here, degrowth theorists and many evolutionary theorists would unite in criticizing the material motivation approach.

The degrowth movement is both creative and constructive when presenting alternative forms of transactions aside from the conventional commodity market. For instance, Fournier mentions 'local trading schemes, gifts, mutual exchange, voluntary contributions', within and between groups (Victor, 2010, p. 371). More hybrid-oriented versions involve renting and sharing, where companies may arrange the schemes (Mont, 2002). Whereas some proponents of the material motivation approach would criticize such alternatives as intrinsically less appealing to people than private ownership, no substantial objections can arguably be found against such alternatives in evolutionary theory (such as evolutionary psychology or anthropology). Extensive anthropological studies and historical studies clearly indicate the plethora of possible alternatives to private ownership and accumulation of goods. Like private ownership, such alternatives may very well be organized so that their dynamics are based on vibrant expressions of our social motivation. In fact, through the inventive, creative, and collaborative component of such alternatives – often exposed to others – these alternatives may perhaps have at least as strong a potential as private ownership on a long-term basis. As to mutual gifts and aid, a large number of studies have indicated the social force of this, given that the organization and setting is one in which such practices may be exposed and efficiently signalled to others (Brace-Govan & Binay, 2010; Dunbar, 2008). The social motivation of 'conspicuous consumption', as Veblen (1899/1967) termed the way we

DOI: 10.1057/9781137276803

show off our wealth and status by purchasing many and expensive goods, should be capable of being at least partially replaced by conspicuous sharing, exchanging, lending, or renting. Still, acting against this strong force of human social motivation we find inertia, short-term interests, and habits of companies as well as citizen-consumers. Yet, increasing numbers of companies are acknowledging the potential of product and service systems. Although mainly at niche levels, increasing numbers of groups and citizens are renting, exchanging, sharing, lending, and purchasing used products from each other and from progressive companies rather than only purchasing all items new in the store (Mont & Tukker, 2006).

Degrowth scholars and practitioners correctly see that mass consumerism and material accumulation are not basic human driving forces. Still, they are sometimes unclear in their views on the nature of the motivation that in current society drives citizen-consumers to accumulate goods, buy new products, and throw away the old ones. Degrowth theorists sometimes express the issue as if competition, comparison, and material envy were motivations constructed by a consumerist society. If we altered society, such constructed motivation could be replaced by the original, truly human character, naturally directed towards ecological motivation and harmony between people. The degrowth vision is formulated as

> a society built on quality rather than on quantity, on cooperation rather than on competition [...] humanity liberated from economism for which social justice is the objective. (Martínez-Alier, Pascual, Vivien, & Zaccai, 2010, p. 1742)

This may indeed sound like a pleasant alternative society for many of us who claim that we do not compete, we only 'work hard' (thus ignoring that all of us in one way or another compete). Yet, as evolutionary psychology, comparative anthropology, and the entire paradigm of New Darwinism has taught us, both competition and cooperation are fundamental parts of human nature and of all human societies. Thus, it is impossible to remove competition from society. What could be done, however, is to make use of the enormous evolutionary force of competition, and develop environmental policies where society's members compete in developing the most harm-reducing practices and the best environmental cooperation.

A question mark should be straightened out concerning a key point of the degrowth ideal, namely the embracing of localism and downscaling. From ecological perspectives, scaling down through increased local food production and material independence, is a controversial strategy.

DOI: 10.1057/9781137276803

Whereas we see strong signs of a social motivation among citizen-consumers to support local food initiatives, there is no scientific consensus among ecologists about what role local production should play on a global scale. Leaving the specific case of local food production aside, the degrowth theorists seem to have an excessive belief in geographical localism as socially motivating to citizen-consumers. To be sure, it is easy to imagine that evolutionary theory and a social motivation approach would proclaim local, small-scale communities similar to those in our ancestral environment. Yet, evolutionary psychology and anthropology have repeatedly indicated how the growth of social networks, often on a global scale, can be fully in line with our social motivation, through the development of communities, sub-communities, and 'weak ties' that transcend all national borders (see Chapter 5). If degrowth theory is to prosper and develop further, the small beauty that it envisions must at the same time connect with the social driving force to broaden the circles far beyond our geographical proximity.

6.4 Conclusions and discussion

Throughout history, people have probably sketched visions for positive, wide-ranging changes to their society. In the fine arts, such visions have usually been evaluated based on aesthetic quality and the level of thought provocation. Visions in social or political writings have been evaluated on the basis of their level of realism. But how can we know what would be a realistic societal vision? In a vague sense, the evaluator runs the risk of putting 'realistic' on a par with 'not too far from the situation we have in society today'. That is clearly the wrong way to separate good visions and scenarios from bad ones, both concerning 'value-free' and normative visions. Instead, this chapter has used as a measuring stick the level of compatibility with lessons from social sciences informed by evolutionary theory. The chapter has analysed three recent green visions of social thinkers: ecological citizenship, alternative hedonism, and degrowth, and commented on these visions based on the extent to which they are coherent with various aspects of (evolutionary based) social motivation.

Alternative hedonism has the advantage of focusing on parts of our human drivers, here called hedonism. Kate Soper argues that what we in consumer society think makes us happy and satisfied does not do so. Soper uses empirical studies, albeit mainly from the humanities and

DOI: 10.1057/9781137276803

much less from all the other relevant disciplines, to support her argument. Yet, she downplays important parts of social motivation, such as competition, the urge to create envy, and social distinction, all of which are crucial to the current appeal of mass consumerism. With the notion of ecological citizenship, Andrew Dobson sketches a vision well grounded in classical political philosophy which many of us would applaud from an intellectual and moral perspective. In the North and West, we are obliged to reduce our ecological footprints to the Southern size. Still, the vision of ecological citizenship downplays the gap between normative ethics (what citizen-consumers should ideally do), and descriptive ethics (what social sciences, particularly with an evolutionary perspective, show empirically and theoretically to be within our motivational reach). Instead, he seems to adopt an ecological motivation approach, which has been shown to be problematic from an evolutionary perspective. Finally, degrowth theorists speak against economism and globalization, instead envisioning highly self-sufficient, small-scale societies. They acknowledge that material accumulation is not fundamental to humans, and they have strong hope that a degrowth society would meet our social needs to a much higher extent than current society. Still, their preoccupation with localism goes against our social motivation to widen our horizon and interact with others on a much wider scale. Moreover, they explicitly disregard the human predisposition for competition and distinction.

Visions and scenarios may function as valuable inspiration and influence on all spheres of society. Yet in order to gain wide cultural resonance, they cannot be based on a blank slate with a random wish list that intelligent scholars have written in any order they like. Instead, the social motivation needs to be taken into account. Thus, global environmental concern can be intellectually sought, but cannot become strong enough drivers for most of us to act upon in isolation from the social motivation. Among ecological visions, those influenced by gender studies of relational goods and reciprocal services seem to be more in line with the social motivation approach than visions based on formalized ethical principles. Overall, the previously mentioned visions and scenarios for wide societal change, along with others, can be much improved if we use the glasses provided by evolution-informed scholarship. I would argue this is one of the few fruitful criteria for assessing the 'realism' of human-oriented visions and scenarios.

DOI: 10.1057/9781137276803

7
Conclusions and Discussion

Abstract: *Chances for 'serious social change' are discussed. Pessimists are right in doubting the end of competition, distinction, envy, and contempt. Optimists are right in doubting that economic greed and turbo consumption are fundamental to human nature. How 'Darwin-free' social science can become compatible with evolutionary theory is briefly elaborated.*

Klintman, Mikael. *Citizen-Consumers and Evolution: Reducing Environmental Harm through Our Social Motivation*. Basingstoke: Palgrave Macmillan, 2013.
DOI: 10.1057/9781137276803.

This book started with a reflection on hope. It described the high hopes – sometimes even optimism – that some scholars and cultural personalities nourish for a new global unity that may take care of threats to the environment. On the other hand, there are, of course, the pessimists. For instance, the philosopher Slavoj Zizek maintains that it is much easier to imagine the end of the world than serious social change (Zizek, 2010).

Can this book be a comment on such grand statements? Possibly. Having examined social motivation, the core human driver, from the lens of social sciences informed by evolutionary theory, I have to admit that Zizek's statement is reasonable in one respect. If by serious social change he is referring to a society where the human traits routinely described as negative in the soft social sciences have been eradicated – competition, envy, status struggles, conflict, and social stratification – he is probably plausible in his doubts. These human traits are biologically inherent in our social motivation. Thus, they will not disappear until – if this ever happens – *Homo sapiens* has evolved further, by 'natural biology' or by genetic modification, into something new. If, however, Zizek' s serious social change refers to a society where various 'stakeholders', including citizen-consumers, have redirected their social motivation into less environmentally harmful practices and preferences, his doubts are less plausible; such 'social changes' are humanly possible, at least. To respond to the other side, namely optimistic assertions about global unity, this book contends that there is no blank slate on which humanity as a whole can write an unconditional wish list, if we aspire to more than utopian thought experiments.

In this book I have developed a strong social motivation approach. The method for doing this has been an examination of obstacles and potentials for everything from environmental adaptation of citizen-consumers' lifestyles, via acceptance of environmental policy and planning, to environmental activism and wider societal change. The book has pointed out the need for acknowledging:

- the fundamental role of a Dionysian (more instinctive and socially oriented) trust in such activities, rather than pinning all hopes on factual environmental information, is one such factor;
- that citizen-consumers cannot be turned into 'environmentally consistent', autonomous individuals driven by ecological motivation is another one;
- that what is often portrayed as 'human greed' and 'manic consumerism' are actually only two among several possible ways

in which we may channel our need to express social status and group belonging. Instead, vanity is a much more appropriate term for the basic trait in our social motivation. Vanity, in turn, is, one must admit, often channelled into practices that in effect are economically greedy and resource-depleting. Importantly, however, our vanity may alternatively be satisfied by various versions of 'green' consumerism, or by what could be labelled 'conspicuous generosity';

- that the sum of economic incentives and disincentives is often overridden by the social context, the decision-making process, and group influence, in environmental policy measures;
- that environmental activism 'from below' will remain fundamental to a change towards far-reaching reductions of environmental harm. The common Western image that 'the institutionalization of environmental issues' in current regulations makes environmental activism redundant is a faulty one, not least from the perspective of evolutionary theory.

Is there any place for normative claims in a book that analyses the evolutionary basis for the social motivation of citizen-consumers? To be sure, this book has criticized the ecological motivation approach for taking for granted, or 'moralizing', that citizen-consumers should turn their environmental awareness into reduced harm as individuals. For example, by using the old philosophical separation of 'is' and 'ought', explained in 2.2.1, and applying it to evolutionary thought, this book criticized the notion of ecological citizenship for implying that an 'ought to' (to reduce environmental harm) would follow logically or psychologically from an 'is' (facts about environmental harms). My criticism does not, of course, contend that citizen-consumers should not have to act upon their awareness of environmental effects of their lifestyles and political support. Instead the problem is that the preceding imperative is 'normative ethics' (about what people ought to do) devoid of 'descriptive ethics' (about what has been seen empirically to be within the reach of people's motivation to do; Booth, 2009). We have evolved biologically into a species that disregards a wide range of 'is', such as long-term consequences of environmental harm, while caring endlessly about a range of social 'is', by struggling for our social position, signalling it, joining groups, distinguishing us from other groups, developing and sharing skills, and so forth. If scholars, policy practitioners, NGOs, and environmental

DOI: 10.1057/9781137276803

pioneers among citizen-consumers are to commit to the naturalistic fallacy of mixing 'is' with 'ought to', it is the 'is' as in social motivation that should be used as the basis for conclusions about what 'ought to' be done.

What would this mean in practice? It would mean that we as citizen-consumers, rather than being expected to dramatically reduce environmental harm as autonomous, anonymous individuals (which we are bad at) instead 'ought to' do what we are good at: discuss and collaborate with friends, neighbours, colleagues, as well as more distant acquaintances, and form or join groups, cooperatives, coalitions, networks, alliances, and NGOs, through which our social motivation may direct us to reduce our environmental harm and to initiate and support changes in policies in a sustainable direction. The moral imperative directed to policy makers and companies would be to stimulate and facilitate such activities, and to create policy sequences that may help to speed up changes in social norms. This book has argued that the latter part, the efforts needed from policy makers and companies, is likely to be more challenging to bring about than the former, group-based efforts by citizen-consumers. To be sure, the citizen-consumer activities entail obstacles such as inert social norms about 'normal', middle-class life, and habits of signalling one's social position through goods and services. Still, I maintain that the social motivation would allow citizen-consumers to turn social signalling and group participation from material accumulation and conspicuous consumption of large material resources into activities with low-resource use. Again, this would presuppose that the low-resource alternatives correspond to our social motivation for competition, signalling, and participation. The efforts needed from the political sphere and industries, however, concern how they should become motivated to invite policy changes beyond a mildly adjusted status quo. Substantial policy changes, both through progressive political initiatives at the top and through adopting changes demanded through the activism of citizen-consumers, presuppose the political and corporate courage to abandon some of the short-term stability that the status quo provides to the high social positions. Yet, those who have this courage may become elevated to an even higher social position, that of heroines and heroes. This social risk situation is part of an inevitable game in cultural evolution (Boyd & Richerson, 2005).

In addition to this book's ambition to analyse current and potential roles for citizen-consumers as regards environmental harm, it is an

DOI: 10.1057/9781137276803

attempt to show the need for the social sciences to become compatible with evolutionary theory. In the same way that it would be highly problematic if biological studies contradicted essential lessons in chemistry, social sciences must be compatible with the essentials of evolutionary theory (Barkow, 2005). The chief problem with the alternative approaches that this book has examined – the ecological motivation approach and the material motivation approach – is that they fail to be compatible with evolutionary theory. Thus, they give an erroneous picture of citizen-consumers and preconditions for environmentally responsible social change. However, I have also tried to show that creating compatibility between evolutionary theory and social science is in many cases doable. Moreover, there are many examples, some illustrated in this book, of full compatibility that seems to be more intuition-based than formally planned. In discussing the compatibility issue between social science and evolutionary theory, I have also had the reverse ambition of indicating how evolutionary theory should learn from compatible studies in social sciences. Many detailed and comprehensive social science studies about group formation, resource mobilization of social movements, and multilevel governance may certainly help develop evolutionary theory further, the latter literature typically having a rather narrow focus on individuals and small groups. The research on citizen-consumers and environmental problems is only one area where increased interaction and compatibility between social sciences and evolutionary scholarship needs to be strengthened. This actually goes for all areas of social science, if it is to prosper and advance its understanding of issues that involve human beings. And as Darwin famously stated,

> 'In the long history of humankind (and animal kind, too) those who learned to collaborate and improvise most effectively have prevailed.'

DOI: 10.1057/9781137276803

References

Abrahamse, W., Steg, L., Vlek, C., & Rothengatter, T. (2005). A review of intervention studies aimed at household energy conservation. *Journal of Environmental Psychology*, 25(3), 273–291.

Ajzen, I., & Fishbein, M. (1980). *Understanding Attitudes and Predicting Social Behavior* (1st ed.). New York: Prentice Hall.

Alcock, J. (2003). *The Triumph of Sociobiology*. Oxford, UK: Oxford University Press.

Andreasen, D. A. R. (2005). *Social Marketing in the 21st Century*. London: Sage Publications, Inc.

Andrews, C. J. (2000). Building a micro foundation for industrial ecology. *Journal of Industrial Ecology*, *4*(3), 35–52.

Atkinson, A. B. (2009). Economics as a moral science. *Economica*, New Series, *76*, 791–804.

Attari, S. Z., Schoen, M., Davidson, C. I., DeKay, M. L., Bruine de Bruin, W., Dawes, R., & Small, M. J. (2009). Preferences for change: do individuals prefer voluntary actions, soft regulations, or hard regulations to decrease fossil fuel consumption? *Ecological Economics*, *68*(6), 1701–1710.

Axelrod, R. M. (1997). *The Complexity of Cooperation: Agent-Based Models of Competition and Collaboration*. Princeton: Princeton University Press.

Baddeley, M. (2009). Herding, social influence and economic decision-making: socio-psychological and neuroscientific analyses. *Philosophical Transactions of the Royal Society B: Biological Sciences*, *365*(1538), 281–290.

Ballantine, P. W., & Creery, S. (2010). The consumption and disposition behaviour of voluntary simplifiers. *Journal of Consumer Behaviour*, 9(1), 45–56.
Balsiger, P. (2010). Making political consumers: the tactical action repertoire of a campaign for clean clothes. *Social Movement Studies*, 9(3), 311–329.
Barkow, J. H. (ed.). (2005). *Missing the Revolution: Darwinism for Social Scientists* (1st ed.). Oxford, UK: Oxford University Press.
Barkow, J. H., Cosmides, L., & Tooby, J. (1992). *The Adapted Mind: Evolutionary Psychology and the Generation of Culture*. Oxford: Oxford University Press.
Barr, S., Shaw, G., Coles, T., & Prillwitz, J. (2010). 'A holiday is a holiday': practicing sustainability, home and away. *Journal of Transport Geography*, *18*(3), 474–481.
Bauman, Z. (1993). *Modernity and Ambivalence* (New ed.). Cambridge, UK: Polity.
Bechtel, R. B., & Churchman, A. (eds). (2002). *Handbook of Environmental Psychology* (1st ed.). New York: Wiley.
Beck, U. (1986/1992). *Risk Society: Towards a New Modernity* (1st ed.). London: Sage Publications Ltd.
Beck, U. (1999). *World Risk Society*. Cambridge, UK: Polity.
Beck, U. (2000). *What Is Globalization*. Cambridge, UK: Polity.
Beck, U. (2011). Cosmopolitanism as imagined communities of global risk. *American Behavioral Scientist*, 55, 1346–1361.
Beckman, L. (2001). Virtue, sustainability and liberal values. In J. Barry & M. Wissenburg (eds), *Sustaining Liberal Democracy: Ecological Challenges and Opportunities* (pp. 279–291). Houndmills, Basingstoke: Palgrave Macmillan.
Bellaby, P. (2010). Uncertainties and risks in transitions to sustainable energy, and the part 'trust'might play in managing them: a comparison with the current pension crisis. *Energy Policy*, *38*, 2624–2630.
Bentley, A., Earls, M., & O'Brien, M. J. (2011). *I'll Have What She's Having: Mapping Social Behavior*. Cambridge, Massachusetts: MIT Press.
Binmore, K. (2005). Economic man – or straw man? *Behavioral and Brain Sciences*, *28*(06), Retrieved from http://www.journals.cambridge.org/abstract_S0140525X05230145
Bok, D. (2010). *The Politics of Happiness: What Government Can Learn from the New Research on Well-Being* (1st ed.). Princeton: Princeton University Press.

Booth, C. (2009). A motivational turn for environmental ethics. *Ethics and the Environment, 14*(1), 53–78.

Boström, M., & Klintman, M. (2009). The green political food consumer. In L. Terragni, M. Boström, B. Halkier, & J. Mäkelä (eds), *Anthropology of Food, Can consumers save the World?*, (S5). Retrieved from http://aof.revues.org/index6394.html

Boström, M., & Klintman, M. (2011). *Eco-Standards, Product Labelling and Green Consumerism*. Houndmills, Basingstoke: Palgrave Macmillan.

Boyd, R., Gintis, H., Bowles, S., & Richerson, P. J. (2003). The evolution of altruistic punishment. *Proceedings of the National Academy of Sciences, 100*(6), 3531–3535.

Boyd, R., & Richerson, P. J. (2005). *The Origin and Evolution of Cultures* (1st ed.). Oxford: Oxford University Press.

Boykoff, M. T., & Goodman, M. K. (2009). Conspicuous redemption? Reflections on the promises and perils of the [`]Celebritization' of climate change. *Geoforum, 40*(3), 395–406.

Brace-Govan, J., & Binay, I. (2010). Consumption of disposed goods for moral identities: a nexus of organization, place, things and consumers. *Journal of Consumer Behaviour, 9*(1), 69–82.

Brody, N. (1983). *Human Motivation: Commentary on Goal-Directed Action*. Academic Pr.

Brown, D. E. (2004). Moral and religious universals as a subset of human universals. *The Science of Good And Evil: Why People Cheat, Gossip, Care, Share, and Follow the Golden Rule*, p. 285.

Brown, D. E. (2004). Human universals, human nature & human culture. *Daedalus, 133*(4), 47–54.

Buss, D. M. (2009). The great struggles of life: Darwin and the emergence of evolutionary psychology. *American Psychologist, 64*(2), 140.

Buttel, F. H. (2003). Environmental sociology and the explanation of environmental reform. *Organization & Environment, 16*(3), 306–344.

Bäckstrand, K., Khan, J., Kronsell, A., & Lövbrand, E. (2010). *Environmental Politics and Deliberative Democracy: Examining the Promise of New Modes of Governance*. Chelterham: Edward Elgar.

Camic, C. (1986). The matter of habit. *American Journal of Sociology, 91*(5), 1039–1087.

Carlsson-Kanyama, A., & Lindén, A. L. (2007). Energy efficiency in residences—challenges for women and men in the North. *Energy policy, 35*(4), 2163–2172.

DOI: 10.1057/9781137276803

Cheshire, T. (2011). Music to think by: Q&A with David Lynch on his debut album. *Wired magazine, 12*. Retrieved from http://www.wired.co.uk/magazine/archive/2011/12/play/music-think-by .

Cohen, M. J. (2010). Destination unknown: pursuing sustainable mobility in the face of rival societal aspirations. *Research Policy, 39*(4), 459–470.

Colombo, R. J. (2009). Exposing the myth of homo economicus (Book Review of 'Moral Markets: The Critical Role of Values in the Economy'). Harvard Journal of Law and Public Policy, 32, 737.

Coolidge, M. L. (1941). Ethics–Apollonian and Dionysian. *The Journal of Philosophy, 38*(17), 449–465.

Cournoyer, D. E. (2004). Evaluating claims for universals: a method analysis approach. *Cross-Cultural Research, 38*(4), 319–342.

Crane, A. (2005). Meeting the ethical gaze: challenges for orienting to the ethical market. In R. Harrison, T. Newholm, & D. Shaw (eds), *The Ethical Consumer* (pp. 219–232). London: Sage.

Cronk, L. (2005). Behavioral ecology and the social sciences. In J.H. Barkow, J. H. (ed.), *Missing the Evolution: Evolutionary Psychology for Social Scientists* (pp. 167–185). Oxford: Oxford University Press.

Daily Mail Online (2008). Eco campaigner Sting flies solo on private jet - twice in one day. (30 June 2008). Retrieved from http://www.dailymail.co.uk/news/article-1030594/Eco-campaigner-Sting-flies-solo-private-jet--twice-day.html

Daley, D. M. (2007). Citizen groups and scientific decisionmaking: does public participation influence environmental outcomes? *Journal of Policy Analysis and Management, 26*(2), 349–368.

Darwin, C. (1879/2004). *The Descent of Man*. London: Penguin Classics.

De Young, R. (2000). Expanding and evaluating motives for environmentally responsible behavior. *Journal of Social Issues, 56*(3), 509–526.

Delingpole, J. (2012, February 18). Eco-loons on the march. *Spectator*, (18 February 2012, www.spectator.co.uk, p. 48).

Devinney, T. M., Auger, P., & Eckhardt, G. M. (2010). *The Myth of the Ethical Consumer*. Cambridge, UK: Cambridge University Press.

Dietz, T., & Stern, P. (2002). *New Tools for Environmental Protection: Education, Information, and Voluntary Measures* (1st ed.). London: National Academies Press.

Dobson, A. (2004). Ecological citizenship. Paper presented at the annual meeting of the Western Political Science Association, Portland, Oregon, USA, March 11, 2004.

DOI: 10.1057/9781137276803

Dobson, A. (2007). Environmental citizenship: towards sustainable development. *Sustainable Development*, *15*(5), 276–285.

Dore, J., & Lebel, L. (2010). Gaining public acceptance: a critical strategic priority of the World Commission on Dams. *Water Alternatives*, *3*(2), 124–141.

Druckman, A., & Jackson, T. (2009). The carbon footprint of UK households 1990–2004: a socio-economically disaggregated, quasi-multi-regional input-output model. *Ecological Economics*, *68*(7), 2066–2077.

Dryzek, J. S. (2011). *Foundations and Frontiers of Deliberative Governance.* Oxford: Oxford University Press.

Dunbar, R. (2008). Showing off in humans: male generosity as a mating signal. *Evolutionary Psychology*, *6*(3), 386–392.

Dunlap, R. E., & York, R. (2008). The globalization of environmental concern and the limits of the postmaterialist values explanation: evidence from four multinational surveys. *The Sociological Quarterly*, *49*(3), 529–563.

Eagly, A. H., & Chaiken, S. (1993). *The Psychology of Attitudes*. Belmont, California: Wadsworth Publishing Co Inc.

Eckersley, R. (2004). *The Green State: Rethinking Democracy and Sovereignty*. Cambridge, Massachusetts: MIT Press.

Ehrlich, P. R., & Ehrlich, A. H. (2008). *The Dominant Animal: Human Evolution and the Environment* (1st ed.). Washington DC: Island Press.

Ek, K., & Söderholm, P. (2008). Norms and economic motivation in the Swedish green electricity market. *Ecological Economics*, *68*(1–2), 169–182.

Ekberg, M. (2007). The parameters of the risk society: a review and exploration. *Current Sociology*, *55*(3), 343–366.

Engel, C., & Singer, W. (2008). *Better Than Conscious? Decision Making, the Human Mind, and Implications for Institutions.* (1st ed.). Cambridge, Massachusetts: MIT Press.

Engelhaupt, E. (2008). Do food miles matter? *Environmental Science & Technology*, *42*(10), 3482–3482.

Eurobarometer (2011). Attitudes of European citizens towards the environment. *Special Eurobarometer*, (365). Retrieved from http://ec.europa.eu/public_opinion/archives/ebs/ebs_365_en.pdf

Fielding, K. S., McDonald, R., & Louis, W. R. (2008). Theory of planned behaviour, identity and intentions to engage in environmental activism. *Journal of Environmental Psychology*, *28*(4), 318–326.

DOI: 10.1057/9781137276803

Fischer, C. S. (2008). Paradoxes of American individualism. *Sociological Forum*, 23(2), 363–372.

Fisher, J., & Irvine, K. (2010). Reducing household energy use and carbon emissions: the potential for promoting significant and durable changes through group participation. Proceedings of conference: IESD PhD Conference: Energy and Sustainable Development Institute of Energy and Sustainable Development, Queens Building, De Montfort University, Leicester, UK, 21 May 2010. Retrieved from http://www.iesd.dmu.ac.uk/events/phd_conference_2010/papers/Fisher.pdf

Fligstein, N., & McAdam, D. (2011). Toward a general theory of strategic action fields. *Sociological Theory*, 29(1), 1–26.

Freeman, D. (1999). *The Fateful Hoaxing of Margaret Mead: A Historical Analysis of Her Samoan Research*. New York: Basic Books.

Freese, J., Li, J.-C. A., & Wade, L. D. (2003). The potential relevances of biology to social inquiry. *Annual Review of Sociology*, 29(1), 233–256.

Freud, S. (1930). *Civilization and Its Discontents*. New York: W. W. Norton & Co.

Frey, B. S. (2000). Motivation and human behaviour. In Peter Taylor-Gooby (ed.), *Risk, Trust and Welfare* (pp. 31–50). Houndmills, Basingstoke: Palgrave Macmillan.

Gangestad, S. W. G., & Simpson, J. A. S. (2007). *The Evolution of Mind: Fundamental Questions and Controversies* (1st ed.). New York: The Guilford Press.

Gaspar, A. (2006). Universals and individuality in facial behavior—past and future of an evolutionary perspective. *Acta Ethologica*, 9(1), 1–14.

Gat, A. (2000). The human motivational complex: evolutionary theory and the causes of hunter-gatherer fighting, part II. Proximate, subordinate, and derivative causes. *Anthropological Quarterly*, 74–88.

Geels, F. W., & Schot, J. (2007). Typology of sociotechnical transition pathways. *Research Policy*, 36(3), 399–417.

Georgescu-Roegen, N. (1975). Energy and economic myths. *Southern Economic Journal*, 41(3), 347–381.

Giddens, A. (1990). *The consequences of modernity*. Palo Alto, California: Stanford University Press.

Giddens, A. (1991). *Modernity and Self-Identity: Self and Society in the Late Modern Age*. Palo Alto, California: Stanford University Press.

Gilg, A., Barr, S., & Ford, N. (2005). Green consumption or sustainable lifestyles? Identifying the sustainable consumer. *Futures*, 37(6), 481–504.

Goodman, J. R., & Sanders, C. R. (2011). In favor of tipping the balance: animal rights activists in defense of residential picketing. *Society and Animals*, *19*(2), 137–155.

Gould, K. A., Pellow, D., & Schnaiberg, A. (2004). Interrogating the treadmill of production: everything you wanted to know about the treadmill but were afraid to ask. *Organization & Environment*, *17*(3), 296–316.

Granovetter, M. (2005). The impact of social structure on economic outcomes. *The Journal of Economic Perspectives*, *19*(1), 33–50.

Greening Lorna A, Greene David L, & Difiglio Carmen. (2000). Energy efficiency and consumption – The rebound effect – A survey. *Energy Policy*, *28*(6–7), 389–401.

Grin, J., Rotmans, J., & Schot, J. (2010). *Transitions to Sustainable Development: New Directions in the Study of Long Term Transformative Change*. Routledge Studies in Sustainability Transitions (1st ed.). London: Routledge.

Griskevicius, V., Tybur, J. M., & Van den Bergh, B. (2010). Going green to be seen: status, reputation, and conspicuous conservation. *Journal of Personality and Social Psychology*, *98*(3), 392–404.

Grolleau, G., Ibanez, L., & Mzoughi, N. (2009). Too much of a good thing? Why altruism can harm the environment? *Ecological Economics*, *68*(7), 2145–2149.

Hammond, K. R. (2007). *Beyond Rationality: The Search for Wisdom in a Troubled Time* (1st ed.). Oxford: Oxford University Press.

Heinonen, S., Jokinen, P., & Kaivo-oja, J. (2001). The ecological transparency of the information society. *Futures*, *33*(3–4), 319–337.

Henrich, J., Boyd, R., Bowles, S., Camerer, C., Fehr, E., Gintis, H., McElreath, R., Alvard, M., Barr, A., Ensminger, J., Hill, K., Gil-White, F., Gurven, M., Marlowe, F., Patton, J., Smith, N., and Tracer, D. (2005). 'Economic man' in cross-cultural perspective: behavioral experiments in 15 small-scale societies. *Behavioral and Brain Sciences*, *28*(6), 795–815.

Hertwich, E. G. (2005). Consumption and the rebound effect: an industrial ecology perspective. *Journal of Industrial Ecology*, *9*(1–2), 85–98.

Hirsch, F. (1999). *Social Limits to Growth*. Bloomington, IN: iUniverse.

Hirsh, J. B., & Dolderman, D. (2007). Personality predictors of consumerism and environmentalism: a preliminary study. *Personality and Individual Differences*, *43*(6), 1583–1593.

DOI: 10.1057/9781137276803

Holzer, B. (2010). *Moralizing the Corporation: Transnational Activism and Corporate Accountability.* Chelterham: Edward Elgar.
Horne, R. E. (2009). Limits to labels: the role of eco-labels in the assessment of product sustainability and routes to sustainable consumption. *International Journal of Consumer Studies, 33*(2), 175–182.
Inglehart, R., & Abramson, P. R. (1999). Measuring postmaterialism. *American Political Science Review, 93*(3), 665–677.
International Energy Agency (2011). *World Energy Outlook 2011 Factsheet.* Paris: International Energy Agency.
Innis, R. E. (2002). *Pragmatism and the Forms of Sense: Language, Perception, Technics.* University Park, Pennsylvania: Penn State University Press.
Jackson, T. (2005). Live better by consuming less? Is there a 'double dividend' in sustainable consumption? *Journal of Industrial Ecology, 9*(1–2), 19–37.
Jagers, S. (2009). In search of the ecological citizen. *Environmental Politics, 18*(1), 18–36.
Johnson, J. A. (2011). The argumentative theory of reasoning applies to scientists and philosophers, too. *The Behavioral and Brain Sciences, 34*(2), 81–2.
Jones, C. R., & Richard Eiser, J. (2010). Understanding [`]local' opposition to wind development in the UK: how big is a backyard? *Energy Policy, 38*(6), 3106–3117.
Kahan, D. M., Peters, E., Wittlin, M., Slovic, P., Ouellette, L. L., Braman, D., & Mandel, G. (2012). The polarizing impact of science literacy and numeracy on perceived climate change risks. *Nature Climate Change.* Retrieved from http://www.nature.com/doifinder/10.1038/nclimate1547
Kahneman, D., Knetsch, J. L., & Thaler, R. H. (1991). Anomalies: the endowment effect, loss aversion, and status quo bias. *The Journal of Economic Perspectives, 5*(1), 193–206.
Kallis, G. (2011). In defence of degrowth. *Ecological Economics, 70*(5), 873–880.
Kilian, B., & Elgström, O. (2010). Still a green leader? The European Union's role in international climate negotiations. *Cooperation and Conflict, 45*(3), 255–273.
Klintman, M. (2000). *Nature and the Social Sciences : Examples from the Electricity and Waste Sectors.* Lund: Department of Sociology, Lund University.

Klintman, M. (2002). Arguments surrounding organic and genetically modified food labelling: a few comparisons. *Journal of Environmental Policy and Planning*, *4*(3), 247–259.

Klintman, M. (2009). Participation in green consumer policies: deliberative democracy under wrong conditions? *Journal of Consumer Policy*, *32*(1), 43–57.

Klintman, M. (2012). Issues of scale in the global accreditation of sustainable tourism schemes: toward harmonized re-embeddedness? *Sustainability: Science, Practice & Policy*, *8*(1). Retrieved from http://sspp.proquest.com/static_content/vol8iss1/1008-040.klintman.pdf

Klintman, M., & Boström, M. (2004). Framings of science and ideology: organic food labelling in the US and Sweden. *Environmental Politics*, *13*(3), 612–634.

Klintman, M., & Boström, M. (2008). Transparency through labelling? Layers of visibility in environmental risk management. In C. Garsten & M. Lindh De Montoya (eds), *Transparency in a New Global Order : Unveiling Organizational Visions* (pp. 179–197). Cheltenham: Edward Elgar.

Klintman, M., & Boström, M. (2011). Political consumerism and the transition towards a more sustainable food regime: looking behind and beyond the organic shelf. In G. Spaargaren, P. Oosterveer, & A. Loeber (eds), *Food Practices in Transition: Changing Food Consumption, Retail and Production in the Age of Reflexive Modernity*, Routledge Studies in Sustainability Transitions. London: Routledge. Retrieved from http://www.routledge.com/books/details/9780415880848/

Klintman, M., & Kronsell, A. (2010). Challenges to legitimacy in food safety governance? The case of the European Food Safety Authority (EFSA). *Journal of European Integration*, *32*(3), 309–327.

Klintman, M., Mårtensson, K., & Johansson, M. (2003). *Bioenergi för Uppvärmning – Hushållens Perspektiv* (No. 2003: 1). Lund: Department of Sociology.

Klintman, M., & Waldo, Å. (2008). *Erfarenheter av Vindkraftsetablering–Förankring, Acceptans och Motstånd* (Research Report No. 5866) (p. 58). Stockholm: Swedish Environmental Protection Agency. Retrieved from http://www.naturvardsverket.se/sv/Nedre meny/Webbokhandeln/ISBN/5800/978-91-620-5866-1/

Koppen, K. V., & Markham, W. T. (2008). Protecting Nature: Organizations and Networks in Europe and the USA. London: Edward Elgar.

DOI: 10.1057/9781137276803

Korten, D. C. (2009). *Agenda for a New Economy: From Phantom Wealth to Real Wealth* (1st ed.). San Fransisco: Berrett-Koehler Publishers.

Kotler, P., Roberto, N., & Lee, N. R. (2002). *Social Marketing: Improving the Quality of Life* (2nd ed.). London: Sage.

Kotler, P., & Zaltman, G. (1971). Social marketing: an approach to planned social change. *The Journal of Marketing*, *35*(3), 3–12.

Kourvetaris, G. A. (1997). The Dionysian and apollonian dimensions of ethnicity: a convergence model. *International Review of Sociology*, *7*(2), 229–237.

Krause, R. M. (2009). Developing conditions for environmentally sustainable consumption: drawing insight from anti-smoking policy. *International Journal of Consumer Studies*, *33*(3), 285–292.

Leonard, T. C. (2009). Origins of the myth of social Darwinism: the ambiguous legacy of Richard Hofstadter's Social Darwinism in American thought. *Journal of Economic Behavior & Organization*, *71*(1), 37–51.

Levett, R. (2003). *A Better Choice of Choice: Quality of Life, Consumption and Economic Growth*. London: Fabian Society.

Liu, X., Wang, C., Shishime, T., & Fujitsuka, T. (2010). Environmental activisms of firm's neighboring residents: an empirical study in China. *Journal of Cleaner Production*, *18*(10), 1001–1008.

Lockwood, M., & Platt, R. (2009). *Green Streets Final Report to British Gas*. Institute for Public Policy Research. Retrieved from www.ippr.org

Logsdon-Conradsen, S. C., & Allred, S. L. (2010). Motherhood and environmental activism: a developmental framework. *Ecopsychology*, 2(3), 141–146.

Lopreato, J. (1990). From social evolutionism to biocultural evolutionism. *Sociological Forum*, 5, 187–212.

Lorek, S., & Fuchs, D. (2011). Strong sustainable consumption governance – precondition for a degrowth path? *Journal of Cleaner Production*. Retrieved from http://www.sciencedirect.com/science/article/pii/S0959652611003088

Lundquist, L. J. (2006). Från Globala Åtaganden till Individuellt Gensvar. (Research Report No. 5547). Stockholm: Swedish Environmental Protection Agency. Retrieved from http://www.naturvardsverket.se/Documents/publikationer/620-5547-X.pdf.

Macnaghten, P. (2003). Embodying the environment in everyday life practices. *Sociological Review*, *51*(1), 63–84.

DOI: 10.1057/9781137276803

Malone, E. L., Dooley, J. J., & Bradbury, J. A. (2010). Moving from misinformation derived from public attitude surveys on carbon dioxide capture and storage towards realistic stakeholder involvement. *International Journal of Greenhouse Gas Control*, 4(2), 419–425.

Marris, C. (2001). Public views on GMOs: deconstructing the myths. *EMBO Reports*, 2(7), 545–548.

Martínez-Alier, J., Pascual, U., Vivien, F. D., & Zaccai, E. (2010). Sustainable de-growth: mapping the context, criticisms and future prospects of an emergent paradigm. *Ecological Economics, 69*(9), 1741–1747.

Matthies, E., Klöckner, C. A., & Preissner, C. L. (2006). Applying a modified moral decision making model to change habitual car use: how can commitment be effective? *Applied Psychology*, 55(1), 91–106.

McLaughlin, P. (2011). Climate change, adaptation, and vulnerability: reconceptualizing societal-environment interaction within a socially constructed adaptive landscape. *Organization & Environment*, 24(3), 269–291.

Mead, G. H. (1938). *Philosophy of the Act*. Chicago: University of Chicago Press.

Mead, M. (1928). *Coming of Age in Samoa: A Psychological Study of Primitive Youth for Western Civilisation.* New York: William Morrow Paperbacks.

Mercier, H., & Landemore, H. E. (2010). Reasoning is for arguing: understanding the successes and failures of deliberation. *Political Psychology*. Retrieved from http://papers.ssrn.com/sol3/papers.cfm?abstract_id=1707029

Mercier, H., & Sperber, D. (2011). Why do humans reason? Arguments for an argumentative theory. *Behavioral and Brain Sciences*, 34, 57–111.

Michaelowa, A. (2005). The German wind energy lobby: how to promote costly technological change successfully. *European Environment*, 15(3), 192–199.

Micheletti, M. (2003). *Political Virtue and Shopping: Individuals, Consumerism, and Collective Action.* Houndmills, Basingstoke: Palgrave Macmillan.

Micheletti, M., Stolle, D., & Follesdal, A. (2003). *Politics, Products, and Markets: Exploring Political Consumerism Past and Present.* Piscataway, New Jersey: Transaction Publishers.

Miegel, F. & Johansson, T. (2002). *Sociology of Culture.* Lund, Sweden: Studentlitteratur (in Swedish).

DOI: 10.1057/9781137276803

Miller, G. (2009). *Spent: Sex, Evolution, and Consumer Behavior*. New York: Viking Adult.

Mitchell, R. E. (2001). Thorstein Veblen: pioneer in environmental sociology. *Organization & Environment*, *14*(4), 389–408.

Mol, A., & Spaargaren, G. (2004). Ecological modernization and consumption: a reply. *Society & Natural Resources*, *17*(3), 261–265.

Mont, O. (2002). Clarifying the concept of product-service system. *Journal of Cleaner Production*, *10*(3), 237–245.

Mont, O., & Tukker, A. (2006). Product-service systems: reviewing achievements and refining the research agenda. *Journal of Cleaner Production*, *14*(17), 1451–1454.

Moore, G. (1903). *Principia Ethica*. Buffalo, New York: Prometheus Books.

Mumford, J., & Gray, D. (2010). Consumer engagement in alternative energy – Can the regulators and suppliers be trusted? *Energy Policy*, *38*(6), 2664–2671.

Neilson, L. A. (2010). Boycott or buycott? Understanding political consumerism. *Journal of Consumer Behaviour*, *9*(3), 214–227.

Newholm, T., & Shaw, D. (2007). Studying the ethical consumer: a review of research. *Journal of Consumer Behaviour*, *6*(5), 253–270.

Newig, J., & Fritsch, O. (2009). Environmental governance: participatory, multi-level – and effective? *Environmental Policy and Governance*, *19*(3), 197–214.

Newson, L., & Richerson, P. (2009). Why do people become modern? A Darwinian explanation. *Population and Development Review*, *35*(1), 117–158.

Nietzsche, F. (1872/2010). *The Birth Of Tragedy*. Whitefish, Montana: Kessinger Publishing, LLC.

Nolan, J. M. (2010). 'An inconvenient truth' increases knowledge, concern, and willingness to reduce greenhouse gases. *Environment and Behavior*, *42*, 643–658.

Nyborg, K., Howarth, R. B., & Brekke, K. A. (2006). Green consumers and public policy: on socially contingent moral motivation. *Resource and Energy Economics*, *28*(4), 351–366.

Nye, M., & Burgess, J. (2008). Promoting durable change in household waste and energy use behaviour. A research report completed for the department for environment, food and rural affairs. School of Environmental Sciences, University of East Anglia, Norwich, UK.

DOI: 10.1057/9781137276803

O'Rourke, D. (2005). Market movements: nongovernmental organization strategies to influence global production and consumption. *Journal of Industrial Ecology*, 9(1–2), 115–128.

Odum, H. T., & Odum, E. C. (2001). *A Prosperous Way Down: Principles and Policies*. Boulder, Colorado: University Press of Colorado.

OECD (2011). CO2 emissions from fuel combustion. *Environment: Key Tables from OECD*, No. 1.

Ogburn, W. F. (1922/1966). *Social Change with Respect to Cultural and Original Nature*. New York: Dell Pub. Co.

Ölander, F., & Thøgersen, J. (1995). Understanding consumer behaviour as prerequisite for environmental protection. *Journal of Consumer Policy*, *18*(4), 345–385.

Pasqualetti, M. J. (2011). Social barriers to renewable energy landscapes. *Geographical Review*, *101*(2), 201–223.

Paterson, M., & Stripple, J. (2010). My space: governing individuals' carbon emissions. *Environment and Planning D: Society and Space*, *28*(2), 341–362.

Peattie, K. (2010). Green consumption: behavior and norms. *Annual Review of Environment and Resources*, *35*, 195–228.

Peattie, K., & Peattie, S. (2009). Social marketing: a pathway to consumption reduction? *Journal of Business Research*, *62*(2), 260–268.

Pellizzoni, L. (2004). Responsibility and environmental governance. *Environmental Politics*, *13*(3), 541–565.

Pepper, S. C. (1958). *The Sources of Value*. Berkeley, California.: University of California Press.

Pierre, B. (2011). Dual-process theories of reasoning: the test of development. *Developmental Review*, *31*(2–3), 151–179.

Pinker, S. (2002). *The Blank Slate: The Modern Denial of Human Nature* (1st ed.). New York: Viking Press.

Pinker, S. (2009). *How the Mind Works*. New York: W. W. Norton & Co.

Ploeger, A. (2010). Evolutionary psychology as a metatheory for the social sciences. *Integral Review*, *6*(3), 164–174.

Power, M. (1999). *The Audit Society: Rituals of Verification* (New Ed.). Oxford: Oxford University Press.

Putnam, R. D. (2001). *Bowling Alone: The Collapse and Revival of American Community* (1st ed.). Clearwater, Florida: Touchstone Books by Simon & Schuster.

Rees, W. (2010). What's blocking sustainability. Human nature, cognition, and denial. *Sustainability: Science, Practice, & Policy*, 6(2), 13–25.

DOI: 10.1057/9781137276803

Reeve, J. (2005). *Understanding Motivation and Emotion* (4th ed.). New York: Wiley.

Reich, W. (2010). Charles Darwin: selfishness and altruism. In C. R. Edling & J. Rydgren (eds), *Sociological Insights of Great Thinkers: Sociology through Literature, Philosophy, and Science* (pp. 133–140). Santa Barbara, California: Praeger.

Ricci, M., Bellaby, P., & Flynn, R. (2010). Engaging the public on paths to sustainable energy: who has to trust whom? *Energy Policy*, *38*(6), 2633–2640.

Richerson, P. J., & Boyd, R. (1999). Complex societies. *Human Nature*, *10*(3), 253–289.

Runciman, W. G. (2008). Forgetting the founders. *The Sociological Review*, *56*(3), 358–369.

Scarce, R. (2005). *Eco-Warriors: Understanding the Radical Environmental Movement,* Updates Edition (2nd ed.). Walnut Creek, California: Left Coast Press.

Schneider, F., Kallis, G., & Martinez-Alier, J. (2010). Crisis or opportunity? Economic degrowth for social equity and ecological sustainability. Introduction to this special issue. *Journal of Cleaner Production*, *18*(6), 511–518.

Schor, J. (2008). Tacking turbo consumption. *Cultural Studies*, *22*(5), 588–598.

Schultz, P. W., Nolan, J. M., Cialdini, R. B., Goldstein, N. J., & Griskevicius, V. (2007). The constructive, destructive, and reconstructive power of social norms. *Psychological Science*, *18*(5), 429–434. doi:10.1111/j.1467-9280.2007.01917.x

Schumacher, E. F. (1973). *Small is Beautiful: Economics as if People Mattered*. New York: Harper Perennial.

Schumacher, I. (2010). Ecolabeling, consumers' preferences and taxation. *Ecological Economics*, *69*(11), 2202–2212.

Schwartz, B. (2005). *The Paradox of Choice: Why More Is Less*. New York: Harper Perennial.

Schwartz, S. H., & Bilsky, W. (1987). Toward a universal psychological structure of human values. *Journal of Personality and Social Psychology*, *53*(3), 550–562.

Seuring, S., Sarkis, J., Müller, M., & Rao, P. (2008). Sustainability and supply chain management – An introduction to the special issue. *Journal of Cleaner Production*, *16*(15), 1545–1551.

DOI: 10.1057/9781137276803

Seyfang, G. (2006). Ecological citizenship and sustainable consumption: examining local organic food networks. *Journal of Rural Studies*, *22*(4), 383–395.

Simmel, G. (1903/2010). The metropolis and mental life. In G. Bridge & S. Watson (eds), *The Blackwell City Reader* (pp. 103–110). New York: Wiley.

Shove, E. (2003). *Comfort, Cleanliness and Convenience: The Social Organization of Normality*. Oxford: Berg Publishers.

Shove, E., Watson, M., Hand, M., & Ingram, J. (2008). *The Design of Everyday Life*. Oxford: Berg Publishers.

Slovic, P. (2001). *Smoking: Risk, Perception and Policy* (1st ed.). London: Sage Publications, Inc.

Smith, A. (1776/1982). *The Wealth of Nations: Books 1–3* (*Penguin Classics*). London: Penguin Classics.

Smith, C. U. M. (2005). Evolutionary neurobiology and aesthetics. *Perspectives in Biology and Medicine*, *48*(1), 17–30.

Soper, K. (2008). Alternative hedonism, cultural theory and the role of aesthetic revisioning. *Cultural Studies*, *22*(5), 567–587.

Sorensen, H. C., Hansen, L. K., Hammarlund, K., & Larsen, J. H. (2002). Experience with and strategies for public involvement in offshore wind projects. *International Journal of Environment and Sustainable Development*, *1*(4), 327–336.

Sorrell, S. (2009). Jevons' paradox revisited: the evidence for backfire from improved energy efficiency. *Energy Policy*, *37*(4), 1456–1469.

Spaargaren, G., & Oosterveer, P. (2010). Citizen-consumers as agents of change in globalizing modernity: the case of sustainable consumption. *Sustainability*, *2*(7), 1887–1908.

Steg, L., & Vlek, C. (2009). Encouraging pro-environmental behaviour: an integrative review and research agenda. *Journal of Environmental Psychology*, *29*(3), 309–317.

Stern, P. C. (2000). New environmental theories: toward a coherent theory of environmentally significant behavior. *Journal of Social Issues*, *56*(3), 407–424.

Stets, J. E., & Burke, P. J. (2000). Identity theory and social identity theory. *Social Psychology Quarterly*, *63*(3), 224–237.

Stock, S. (2008). Fear of fallowing: the specter of a no-growth world. *Harper's Magazine*. Retrieved from http://www.harpers. org/ archive/2008/03/0081958

DOI: 10.1057/9781137276803

Sturgis, P., & Allum, N. (2004). Science in society: re-evaluating the deficit model of public attitudes. *Public Understanding of Science*, *13*(1), 55–74.

Sturgis, P., Cooper, H., & Fife-schaw, C. (2005). Attitudes to biotechnology: estimating the opinions of a better-informed public. *New Genetics and Society*, *24*(1), 31–56.

Symons, D. (1995). On the use and misuse of darwin in the study of human behavior. In J. H. Barkow, L. Cosmides, & J. Tooby (eds), *The Adapted Mind: Evolutionary Psychology and the Generation of Culture* (pp. 137–159). Oxford University Press, USA.

Tamm Hallström, K., & Boström, M. (2010). *Transnational Multi-Stakeholder Standardization: Organizing Fragile Non-State Authority.* Edward Elgar Pub.

Taylor-Gooby, P. (2006). *The Efficiency/Trust Dilemma Taylor-Gooby.* University of Kent at Canterbury: Economic & Social Research Council.

Thaler, R. H., & Sunstein, P. C. R. (2008). *Nudge: Improving Decisions About Health, Wealth, and Happiness* (1st ed.). New Haven: Yale University Press.

The Economist (2009, February 28). A load of rubbish. *Economist*, p. 16.

Thøgersen, J., & Crompton, T. (2009). Simple and painless? The limitations of spillover in environmental campaigning. *Journal of Consumer Policy*, *32*(2), 141–163.

Toennies, F. (1887). *Community and Society:* New Brunswick, New Jersey: Transaction Publishers.

Trentmann, F. (2007). Citizenship and consumption. *Journal of Consumer Culture*, *7*(2), 147–158.

Trivers, R. (2011). *Deceit and Self-Deception: Fooling Yourself the Better to Fool Others.* New York: Penguin.

Trivers, R. L. (1971). The evolution of reciprocal altruism. *Quarterly Review of Biology*, *46*(1), 35–57.

Turaga, R. M. ., Howarth, R. B., & Borsuk, M. E. (2010). Pro-environmental behavior. *Annals of the New York Academy of Sciences*, *1185*(Ecological Economics Reviews), 211–224.

Turner, J. H. (1987). Toward a sociological theory of motivation. *American Sociological Review*, *52*(1), 15–27.

Urry, J. (2010). Consuming the planet to excess. *Theory Culture Society*, *27*(2–3), 191–212.

Van Den Bergh, J. C. . (2011). Environment versus growth—A criticism of 'degrowth' and a plea for 'a-growth'. *Ecological Economics*, *70*(5), 881–890.

DOI: 10.1057/9781137276803

Van Der Heijden, H. A. (2006). Globalization, environmental movements, and international political opportunity structures. *Organization & Environment*, *19*(1), 28–45.

Veblen, T. (1899/1967). *The Theory of the Leisure Class: An Economic Study of Institutions.* New York: Funk & Wagnalls.

Victor, P. (2010). Questioning economic growth. *Nature*, *468*(7322), 370–371.

Vliet, B. van, Spaargaren, G., & Oosterveer, P. (eds). (2010). *Social Perspectives on the Sanitation Challenge* (1st ed.). New York: Springer.

Wackernagel, M., & Rees, W. E. (1996). *Our Ecological Footprint: Reducing Human Impact on the Earth.* Gabriola Island, BC Canada: New Society Publishers.

Wadud, Z., Graham, D. J., & Noland, R. B. (2009). Modelling fuel demand for different socio-economic groups. *Applied Energy*, *86*(12), 2740–2749.

Waldo, Å. (2012). Offshore wind power in Sweden—A qualitative analysis of attitudes with particular focus on opponents. *Energy Policy*, *41*, 692–702.

Waldo, Å., & Klintman, M. (2010). *Attityder och Delaktighet vid Etablering av Vindkraft till Havs* (No. 2010–4) (p. 94). Stockholm: Swedish Environmental Protection Agency. Retrieved from http://www.naturvardsverket.se/Documents/publikationer/978–91–620–6351–1.pdf

Warde, A. (1997). *Consumption, Food and Taste: Culinary Antinomies and Commodity Culture.* London: Sage.

Warde, A., & Martens, L. (2000). *Eating Out: Social Differentiation, Consumption, and Pleasure.* Cambridge, UK: Cambridge University Press.

Wauters, E., Bielders, C., Poesen, J., Govers, G., & Mathijs, E. (2010). Adoption of soil conservation practices in Belgium: an examination of the theory of planned behaviour in the agri-environmental domain. *Land Use Policy*, *27*(1), 86–94.

Whitmarsh, L. (2009). Behavioural responses to climate change: asymmetry of intentions and impacts. *Journal of Environmental Psychology*, *29*(1), 13–23.

Wilson, D., & Dixon, W. (2008). Homo economicus meets GH Mead: a contribution to the critique of economic theory. *American Journal of Economics and Sociology*, *67*(2), 241–263.

Wilson, E. O. (2004). *On Human Nature: Revised Edition* (Revised ed.). Cambridge, Massachusetts: Harvard University Press.

DOI: 10.1057/9781137276803

Wilson, E. O. (2012). *The Social Conquest of Earth* (1st ed.). New York: W. W. Norton & Co.

Winslott-Hiselius, L., Brundell-Freij, K., Vagland, Å., & Byström, C. (2009). The development of public attitudes towards the Stockholm congestion trial. *Transportation Research Part A: Policy and Practice*, *43*(3), 269–282.

Wolf, J., Brown, K., & Conway, D. (2009). Ecological citizenship and climate change: perceptions and practice. *Environmental Politics*, *18*(4), 503–521.

Wolsink, M. (2010). Near-shore wind power – Protected seascapes, environmentalists' attitudes, and the technocratic planning perspective. *Land Use Policy*, *27*(2), 195–203.

Wright, R. (1995). *The Moral Animal: Why We Are, the Way We Are: The New Science of Evolutionary Psychology* (1st ed.). New York: Vintage.

Wright, R. (2001). *Nonzero: The Logic of Human Destiny*. New York: Vintage.

Wüstenhagen, R., Wolsink, M., & Bürer, M. J. (2007). Social acceptance of renewable energy innovation: an introduction to the concept. *Energy Policy*, *35*(5), 2683–2691.

Zizek, S. (2010). *Living in the End Times*. New York: Verso.

DOI: 10.1057/9781137276803

Index

DOI: 10.1057/9781137276803

DOI: 10.1057/9781137276803

DOI: 10.1057/9781137276803

DOI: 10.1057/9781137276803

Lightning Source UK Ltd.
Milton Keynes UK
UKOW040407031212

203072UK00002B/4/P

9 781137 276797